Aligning Policies for a Low-carbon Economy

This work is published under the responsibility of the Secretary-General of the OECD. The opinions expressed and arguments employed herein do not necessarily reflect the official views of the member countries of the OECD, the OECD Nuclear Energy Agency, the International Energy Agency or the International Transport Forum.

This document and any map included herein are without prejudice to the status of or sovereignty over any territory, to the delimitation of international frontiers and boundaries and to the name of any territory, city or area.

Please cite this publication as:
OECD/IEA/NEA/ITF (2015), *Aligning Policies for a Low-carbon Economy*, OECD Publishing, Paris.
http://dx.doi.org/10.1787/9789264233294-en

ISBN 978-92-64-23326-3 (print)
ISBN 978-92-64-23329-4 (PDF)

The statistical data for Israel are supplied b180y and under the responsibility of the relevant Israeli authorities. The use of such data by the OECD is without prejudice to the status of the Golan Heights, East Jerusalem and Israeli settlements in the West Bank under the terms of international law.

Photo credits: Cover © Baseline Arts Ltd, Oxford.

Corrigenda to OECD publications may be found on line at: *www.oecd.org/about/publishing/corrigenda.htm*.

Foreword

At the Ministerial Council Meeting in May 2014, ministers and representatives of OECD member countries and the European Union invited the OECD to work with the International Energy Agency (IEA), the International Transport Forum (ITF) and the Nuclear Energy Agency (NEA) "to continue to support the UNFCCC negotiations and to examine how to better align policies across different areas for a successful economic transition of all countries to sustainable low-carbon and climate-resilient economies and report to the 2015 OECD Ministerial Council Meeting." These areas include economic, fiscal, financial, competition, employment, social, environmental, energy, investment, trade, development co-operation, innovation, agriculture and sustainable food production, regional as well as urban, and transport policies.

This report on *Aligning Policies for a Low-carbon Economy* responds to that request by identifying where existing policy and regulatory frameworks are at odds with climate policy, i.e. where existing policies may make climate policy less effective than it could be otherwise. It reflects the initial diagnosis on where and how existing policy and regulatory frameworks may not be aligned with a low-carbon economy.

Acknowledgements

This report is the product of a collaborative effort between the Organisation for Economic Co-operation and Development (OECD), the International Energy Agency (IEA), the Nuclear Energy Agency (NEA) and the International Transport Forum (ITF).

It has been overseen by OECD Deputy Secretary-General Rintaro Tamaki and co-ordinated by Richard Baron, Project Co-ordinator, and Virginie Marchal. Amelia Smith provided continuous editorial and secretarial support. The team is grateful to Simon Upton, for his very useful guidance and leadership throughout the project, to Andrew Prag who drafted the chapter on trade and provided valuable advice on the rest of the report, and to Michael Mullan for the chapter on adaptation.

The authors would also like to thank particularly the following colleagues across the four organisations for their input throughout the project: Anthony Cox, Nick Johnstone, Tomasz Kozluk, Alexis Robert, Ron Steenblik and Kurt Van Dender (all from the OECD), Philippe Benoit (IEA), Jan Keppler (NEA) and Stephen Perkins (ITF). The team is also indebted to Sarah Finke for guiding the publication through the OECD approval process. Government officials from different ministries across OECD countries, and stakeholder representatives, in particular business (co-ordinated through the Business and Industry Advisory Committee of the OECD), and trade unions (co-ordinated through the Trade Union Advisory Committee to the OECD) provided extensive feedback on the first draft of the report during and after a consultation seminar organised in February 2015 at OECD headquarters.

The main authors of *Aligning Policies for a Low-carbon Economy* were:

Chapters	Main authors
Executive Summary	Richard Baron and Virginie Marchal
Chapter 1: Core climate policies and the case for policy alignment	Richard Baron
Chapter 2: Scaling-up low-carbon investment and finance	Virginie Marchal
Chapter 3: Implementing climate-friendly taxation practices	Richard Baron with Žiga Žarnic
Chapter 4: Delivering innovation and skills for the low-carbon transition	Richard Baron and Nick Johnstone
Chapter 5: Removing international trade barriers	Andrew Prag
Chapter 6: Diagnosing misalignments for a more resilient future	Michael Mullan
Chapter 7: Reframing investment signals and incentives in electricity	Richard Baron
Chapter 8: Opting for low-carbon urban mobility	Virginie Marchal
Chapter 9: Strengthening incentives for sustainable land use	Virginie Marchal

Many other colleagues provided comments and suggestions and their contributions are gratefully acknowledged. They include: from the Centre for Entrepreneurship, SMEs and Local Development: Nathalie Cliquot; from the Centre for Tax Policy and Administration: Johanna Arlinghaus; Florens Flues, Michelle Harding; from the Development Co-operation Directorate: Jan Corfee-Morlot and Stephanie Ockenden; from the Directorate for Financial and Enterprise Affairs: Karim Dahou, Raffaele Della Croce, Despina Pachnou, Cristina Tebar-Less, David Pinkus, Joel Paula, Gert Wehinger; from the Economics Department: Silvia Albrizio, Boris Cournède, Alain De Serres, Balázs Egert, Christian Kastrop, Giuseppe Nicoletti and Mauro Pisu; from the Directorate for Employment, Labour and Social Affairs: Alessandro Goglio and Paul Swaim; from the Environment Directorate: Shardul Agrawala, Geraldine Ang, Nils Axel Braathen, Gregory Briner, Simon Buckle, Jean Chateau, Christopher Kaminker, Kumi Kitamori, Nicolina Lamhauge, Walid Oueslati, Robert Youngman and Žiga Žarnic, who also helped on the organisation of an international seminar dedicated to this project; from the International Energy Agency: Heymi Bahar, Manuel Baritaud, Adam Brown, Pierpaolo Cazzola, Laura Cozzi, Davide D'Ambrosio, John Dulac, Anselm Eisentraut, Paolo Frankl, Jean-François Gagné, Timur Guel, Christina Hood, Fabian Kesicki, Ellina Levina, Simon Mueller, Cédric Philibert, Sylvia Spruck, Laszlo Varro and Matthew Wittenstein; from the International Transport Forum: Aimée Aguilar-Jaber, Olaf Merk and José Viegas; from the Nuclear Energy Agency: Thierry Dujardin; from the Public Governance and Territorial Development Directorate: Rudiger Ahrend, Rolf Alter, Charles Baubion, Andrew Davies, Claire Charbit, Isabelle Chatry, Ronnie Downes, Catherine Gamper, Daniela Glocker, Stéphane Jacobzone, Céline Kauffmann, Karen Maguire, Tadashi Matsumoto, Marissa Plouin, William Tompson and Paolo Veneri; from the Directorate for Science, Technology and Innovation: Mario Cervantes, Colin Webb, Kirsten Wiebe, Andrew Wyckoff, Norihiko Yamano; from the Trade and Agriculture Directorate: Ken Ash, Morvarid Bagherzadeh, Jonathan Brooks, Dimitris Diakosavvas, Jared Greenville, Guillaume Gruère, Julien Hardelin, Ada Ignaciuk, Franck Jésus, Jean Le Cocguic, Jehan Sauvage, Frank Van Tongeren and Martin Von Lampe. The report also benefited from comments of external reviewers: Michael Jacobs and James Rydge (New Climate Economy), and Ian Cochran (CDC Climat Recherche).

Fiona Hinchcliffe and Jennifer Allain provided editorial support, and Janine Treves, Katie Kraig-Ernandes, Catherine Bremer, Shayne MacLachlan and Stéphanie Simonin-Edwards supported the publication and communication processes.

Japan and Korea provided financial contributions to the project.

Table of contents

Tables

Figures

Acronyms and abbreviations

2DS	Energy scenario of the International Energy Agency's *Energy Technology Perspectives* compatible with limiting global warming to 2°C
4DS	Energy scenario compatible with limiting global warming to 4°C. See 2DS.
6DS	Energy scenario compatible with limiting global warming to 6°C. See 2DS.
APEC	Asia-Pacific Economic Cooperation
BRT	Bus rapid transit
CBA	Cost-benefit analysis
CCS	Carbon capture and storage
CDM	Clean Development Mechanism
CfD	Contract for Difference (United Kingdom)
CHP	Combined heat and power
CIS	Commonwealth of Independent States
COP21	21st Session of the Conference of the Parties to the United Nations Framework Convention on Climate Change
DAC	Development Assistance Committee (OECD)
DSM	Demand-side management
DSR	Demand-side response
EBRD	European Bank for Reconstruction and Development
ECA	Export credit agency
EPA	Environmental Protection Agency (United States)
ESG	Environmental, social and governance
ETS	Emissions Trading System
EU	European Union
FiT	Feed-in tariff
FSB	Financial Stability Board
GATS	General Agreement on Trade in Services
GATT	General Agreement on Tariffs and Trade
GDP	Gross domestic product

GHG	Greenhouse gas
GPP	Green public procurement
GVC	Global value chain
GW	Gigawatts
ICAO	International Civil Aviation Organization
ICT	Information and communications technology
IEA	International Energy Agency
IMO	International Maritime Organization
INDC	Intended nationally determined contribution
IPCC	Intergovernmental Panel on Climate Change
ITF	International Transport Forum
LCR	Local-content requirement
LTA	Land Transport Authority (Singapore)
MPO	Metropolitan planning organisation
MSW	Municipal solid waste
NDRC	National Development and Reform Commission (People's Republic of China)
NEA	Nuclear Energy Agency
NTB	Non-tariff barrier
ODA	Official development assistance
PDU	Urban mobility plan French: *Plan de déplacements urbains*
PES	Payment for ecosystem services
PFI	Policy Framework for Investment
PFIA	Policy Framework for Investment in Agriculture
PMR	Product market regulation
PPP	Public-private partnership
PRI	Public research institution
PSE	Producer Support Estimate
PV	Photovoltaics
RCP	Representative concentration pathway
RD&D	Research, development and demonstration
REDD+	Reduction in Emissions from Deforestation and Forest Degradation
RMV	Rhein-Main Transport Association (Germany) *Rhein-Main Verkehrsverbund*

RTA	Regional trade agreement
SCM	Agreement on Subsidies and Countervailing Measures
SME	Small and medium-sized enterprise
SNG	Synthetic natural gas
SOE	State-owned enterprise
STRI	Services Trade Restrictiveness Index
T&D	Transmission and distribution
TBT	Agreement on Technical Barriers to Trade
tCO_2-eq	A measure of greenhouse gas emissions, expressed in tonnes of CO_2 equivalent
TfL	Transport for London
TISA	Trade in Services Agreement
TRIMS	Agreement on Trade-Related Investment Measures
UNFCCC	United Nations Framework Convention on Climate Change
VAT	Value-added tax
VET	Vocational education and training
VRE	Variable renewable energy
WTO	World Trade Organization

Preface

Addressing human-induced climate change is one of the most significant challenges to be undertaken by the international community. The problem has long been identified but emissions of greenhouse gases keep rising, and the urgency of action increases with every passing year. Protecting the earth's climate implies a transformational agenda that needs a resolute and enduring commitment. The IPCC Fifth Assessment Report tells us that we need to return global greenhouse gas emissions to a net zero level by the end of the century. That means a drastic departure from the continuous growth in emissions we have witnessed in recent years: global emissions have to peak in the coming decade in order to avoid a global temperature increase above 2°C.

More and more countries are implementing policies to reduce their emissions, including carbon pricing mechanisms and energy efficiency measures. Governments have also actively encouraged the research, development and deployment of low-carbon technologies. These new technologies are starting to make inroads into the share of fossil fuels, even if the latter still account for more than 80% of total energy supply globally. A well-balanced package of carbon pricing, energy efficiency measures and targeted support to low-carbon technologies ought to be at the core of climate mitigation policy. Their implementation needs broadening and deepening to redirect investment away from fossil fuels and towards a low-carbon energy mix.

In 2014, OECD Ministers encouraged our four organisations to work together to answer a fundamental question: are policy frameworks, in areas as diverse as investment, taxation, energy, labour, agriculture and others, well aligned with the pursuit of the transition to a low-carbon, climate-resilient economy? And if the answer is no, can they be better aligned and become new levers for the decarbonisation of the economy? Our joint work started from the observation that the world economy has grown for well over a century on the back of convenient fossil sources of energy, and that moving away from them is a formidable structural challenge. Coal, oil and natural gas have influenced energy systems, the organisation of our cities and transport networks, and how governments raise taxes. The regulatory wiring of our economies was designed for a world powered by fossil fuels long before we identified the impact of rising greenhouse gas emissions on the Earth's climate.

This report confirms that a number of policies are not well aligned with climate policy objectives and are in some cases in direct conflict. Fiscal systems contain a number of provisions that guide consumers and companies *towards* higher fossil energy consumption and production. The broad environment for investment is also not always conducive to investment in the long-term infrastructure needed to fight climate change. Corporate disclosures still lack transparency on companies' exposure to climate risks. Trade measures can also restrict access to low-carbon technologies, increasing the cost of the transition. The policy environment for innovation is not always conducive to the emergence of new technologies and business models.

We have also studied specific activities that are central to the decarbonisation challenge: electricity systems, where market design fails to provide signals for investment in the needed low-carbon, capital-intensive power generation technologies (such as solar, wind, nuclear, and power plants fitted with carbon capture and storage); urban mobility, where integration of land-use planning and sustainable transport systems remains difficult; and rural land use, where too many environmentally harmful subsidies remain, to name just a few examples. In these cases as well, policy frameworks can be revisited to ease the transition to low carbon and this report documents actual policy solutions in developed and developing countries alike.

The sum effect of policy misalignments in an economy can significantly undermine the effectiveness of climate policy efforts. It is important that governments make their own diagnosis of policy frameworks that inadvertently hamper decarbonisation to see what reforms are warranted. All parts of governments, including those outside the usual climate policy portfolio, can contribute. In many cases, solving misalignments can deliver on other important policy objectives: more infrastructure investment, a more inclusive growth, higher energy security and a healthier environment.

Limiting global climate change to within safe boundaries is essential. Effective action requires a better alignment of policy frameworks with climate goals. Our joint report shows governments where they should start looking for more policy levers if they want to ease the transition to a low-carbon economy.

Angel Gurría
Secretary General, OECD

Maria van der Hoven
Executive Director, IEA

José Viegas
Secretary General, ITF

William D. Magwood, IV
Director General, NEA

Executive summary

Addressing climate change requires urgent policy action to drive an unprecedented global infrastructure and technological transformation. More countries are implementing core climate policies: carbon pricing and market-based instruments, regulatory intervention and targeted support to innovation in low-carbon sustainable technologies. But global greenhouse gas emissions have risen rapidly and remain too high to avoid severe and irreversible climate change impacts.

A number of obstacles stand in the way of effective climate policy. Among the most important is the fact that existing policy frameworks and economic interests continue to be geared towards fossil fuels and carbon-intensive activities, as coal, oil and natural gas have fuelled global economic development for centuries. Inadvertently or not, this creates a misalignment between existing policy frameworks and climate objectives, hindering low-carbon investment and consumption choices.

This report presents the first broad diagnosis of misalignments with climate goals in areas essential to the transition to a low-carbon economy. It points to a number of misalignments in policy domains, such as finance, taxation, trade policies, innovation and adaptation, as well as in three specific sectors: electricity, urban mobility, and land-use.

Beyond facilitating climate action, aligning these policies with a low-carbon economy can contribute to a broader reform agenda for greener, more resilient and inclusive growth, including more progressive tax codes, pro-growth long-term infrastructure investment, and energy and transport systems that support cleaner air, better health and a more diversified energy supply.

Better policy alignment for a better climate and better growth

Scale up sustainable low-carbon investment and finance. There is an urgent and unprecedented opportunity to ensure that new investment in infrastructure supports the climate agenda while fostering economic development. The additional short-term costs of shifting to low carbon would amount to just a fraction of the finance needed for infrastructure overall. There is no shortage of capital, but new sources of financing need to be mobilised. Financial stability is a prerequisite to any kind of investment, including low carbon. However, financial regulations could unintendedly limit the supply of long-term finance. Addressing the potential impact of existing financial sector rules could unlock investment in low-carbon infrastructure. Public finance and investment can also catalyse the low-carbon transition provided that governments reconsider their support for investments in greenhouse-gas-intensive activities, and mainstream climate objectives into public procurement and official development assistance.

Look at taxation beyond energy alone. Subsidies and tax expenditures favouring the production and use of fossil fuels slow down low-carbon innovation; however, current low oil prices also present an opportunity for reform. Other taxes and tax provisions deserve a closer look (e.g. property taxes, various corporate income tax provisions), as they may encourage carbon-intensive choices. For example, the tax treatment of company cars encourages more CO_2 emissions across OECD countries. Governments also need to anticipate the impact of the low-carbon transition on tax revenues.

Spur low-carbon innovation on a large scale. Clear and credible government commitment to ambitious core climate policy instruments is an important spur for low-carbon innovation. The low-carbon transition could – and in some cases already is – driving a boom in innovation and emerging businesses, and a parallel shift in skills and the labour force. Innovation for the low-carbon transition is about the creation of new businesses, the restructuring or the phasing out of old ones, the emergence of nascent technologies and business models, and the right support frameworks for innovations to be widely adopted. This includes addressing potential skills gaps through education, training and labour market policies.

Promote climate-friendly international trade. The international trade regime itself does not prevent governments from pursuing ambitious climate policies, but some international trade barriers can undermine climate objectives. For example, import tariffs still penalise trade in some technologies needed for the low-carbon transition. An Environmental Goods Agreement, currently under negotiation, would help to reduce the costs of climate mitigation efforts, among other outcomes. Care needs to be taken by the many countries that are promoting greener growth by favouring domestic manufacturers of low-carbon technologies. Where these measures restrict international trade, they may well undermine overall investment and the uptake of sustainable technologies.

Revisit electricity markets. Electricity lies at the heart of a successful decarbonisation of energy systems. However, deregulated electricity markets do not deliver the long-term price signal needed for investment in high capital cost, low-carbon technologies. Ensuring competitive and timely investment in low-carbon solutions will require new market arrangements such as long-term supply agreements, as well as a robust and stable CO_2 price signal. Jurisdictions with regulated systems that consider introducing greater competition need to adopt market arrangements that will encourage, rather than hinder, investment in low-carbon technologies.

Opt for sustainable urban mobility. Current transport systems, which rely largely on fossil fuels, impose very high environmental costs (climate change, noise, air pollution), particularly in urban settings. Policy intervention is needed to provide more energy efficient and less carbon-intensive mobility. In many cities, land-use and transport planning are poorly co-ordinated and encourage greater use of private cars. Aligning policy action across levels of governments and between stakeholders could do much to deliver lower-carbon mobility. National frameworks and legislation sometimes leave local governments with little financial or political leeway to make low-carbon choices.

Strengthen incentives for sustainable land use. Sustainable land-management practices – reduced deforestation, restoring degraded land, low-carbon agricultural practices and increased carbon sequestration in soils and forests – can make a large contribution to reducing greenhouse gas emissions while responding to growing food demands. They could also improve the resilience of our economies to a changing climate by protecting ecosystems. This requires an integrated approach that breaks down the silos between mitigation, adaptation, agriculture, food security, forestry and environmental policies. More specifically, countries could pursue their efforts to remove environmentally harmful agricultural subsidies, value ecosystem services, protect forests and minimise food waste.

Engaging the low-carbon transition

Climate policy can be more effective if all government ministries identify important misalignments with low-carbon transition in their respective portfolios. An ambitious climate action plan will therefore need new approaches to policy making across government.

Beyond the national level, better alignment of policies across countries could also boost effectiveness and alleviate concerns about potential distortions of competition. A global agreement on greenhouse gas reductions would send a strong signal in this direction.

Chapter 1

Core climate policies and the case for policy alignment

This chapter presents the scientific basis for climate action and the transformative nature of climate policy objectives. Action to mitigate climate change must rest on three pillars: an explicit or implicit price on CO_2 emissions, regulations to remove barriers to energy efficiency, and targeted support to bring low-carbon technologies to market. The chapter highlights the need for stakeholder (consumers, industry) buy-in of these core climate policies, careful consideration by governments, and alignment of broader policy frameworks, traditionally hard-wired to fossil fuels, towards a low-carbon economy. Identifying and reforming misaligned policies can also help the transition while also supporting other policy objectives.

The statistical data for Israel are supplied by and under the responsibility of the relevant Israeli authorities. The use of such data by the OECD is without prejudice to the status of the Golan Heights, East Jerusalem and Israeli settlements in the West Bank under the terms of international law.

Key messages

Policy makers face a number of structural economic, financial, social and environmental challenges. Responding to the threat of climate change implies a profound transformation in order to reduce greenhouse gas emissions over this century, in particular CO_2 from the production and consumption of fossil fuels.

The policy framework required to orient the economy away from greenhouse gas-intensive activities rests on three pillars: 1) sending a robust and credible price signal to internalise the cost of these emissions; 2) regulatory measures whenever pricing is not effective; 3) bringing promising low-emission technologies to commercial maturity in anticipation of more ambitious reductions. More countries are taking action, but implementation is lagging behind in many regions, as the policy changes required tend to be resisted on social and economic grounds.

Some solutions to enhance the acceptability of climate policies are in the hands of domestic policy makers and can serve other policy objectives. For instance, compensation systems could more than offset the distributional impacts on low-income households, and address social concerns. Solving possible distortions in competition stemming from differences in domestic climate policies hinges on international collaboration. A global climate change agreement under the United Nations Framework Convention on Climate Change (UNFCCC) could open the door to this discussion.

A key obstacle to the effectiveness and acceptability of core climate policies is the number of regulatory and policy frameworks outside the climate policy portfolio that are not aligned with climate objectives. Identifying and addressing these misalignments systematically in each country will enhance the responsiveness of economic and social systems to the climate change agenda. In developed and developing countries alike, this approach could help governments be more ambitious in policy implementation and emissions objectives while also achieving greener, more robust and inclusive growth.

This volume presents the first broad diagnosis of the coherence between overall policy and regulatory frameworks and climate goals. It identifies a number of opportunities for realigning policies to enable an efficient and cost-effective shift to a low-carbon economy, across four policy domains (investment, taxation, innovation and skills, and trade) and three specific areas that are important for the low-carbon transition (electricity, urban mobility and rural land use). Adaptation to ongoing climate change can also be enhanced through better alignment. This agenda will require the engagement of parts of government not usually mobilised in the development of climate change response strategies.

Facilitating the low-carbon transition

Addressing climate change and other structural challenges

Addressing climate change is a transformational agenda. Emissions of greenhouse gases (GHG), and in particular CO_2 from fossil fuels, are embedded in the vast majority of human activities. Carbon-intensive sources have fuelled global economic development and will continue to do so for some time. Returning global emissions of GHG to net zero by the end of the century, a requirement for keeping the global average temperature increase below 2°C, will mean profound changes to infrastructures, technologies and behaviours. Failure to do so will increase the likelihood of severe, pervasive and irreversible impacts for human activities and ecosystems, threaten water security, the livelihoods of poor people, agricultural productivity and global economic growth (OECD, 2013a, 2012; and Box 1.1).

Box 1.1. **The scientific basis**

The Intergovernmental Panel on Climate Change's (IPCC) latest projections indicate that the emissions pathways consistent with limiting warming to below 2°C relative to pre-industrial levels require substantial emissions reductions over the next few decades, and near zero emissions of carbon dioxide (CO_2) and other long-lived greenhouse gases (GHGs) by the end of this century (Figure 1.1). The remaining GHG emissions from burning fossil fuels and other activities (e.g. agriculture) will have to be offset. With present technological knowledge, this implies the generation of energy from bioenergy in plants fitted with carbon capture and storage in order to remove CO_2 from the atmosphere.[1]

Figure 1.1. **Greenhouse gas emissions pathways 2000-2100: All scenarios from the IPCC Fifth Assessment Report**

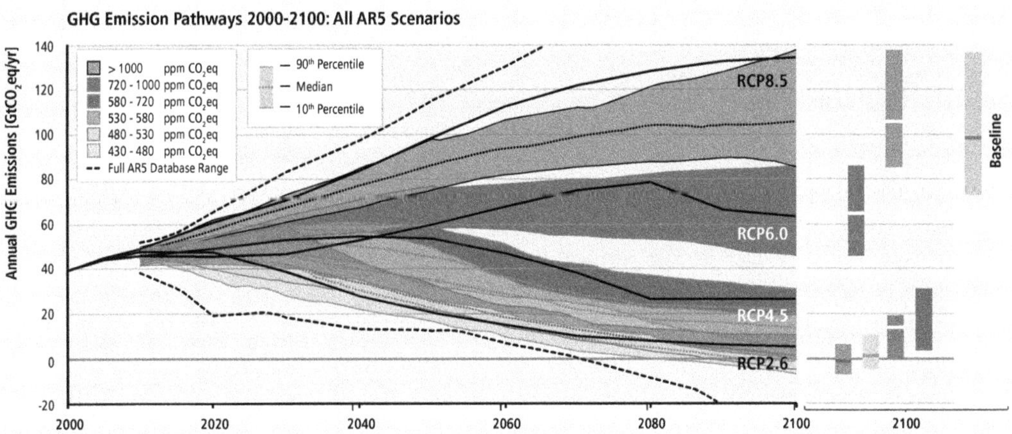

Note: This figure presents all scenarios from the IPCC Fifth Assessment Report, including GHG representative concentration pathways (RCP). Scenarios in the lowest light-blue band correspond to concentration ranges of 430-480 ppm, likely to keep temperatures below 2°C by the end of the century; scenarios in the lowest darker blue band, correspond to concentration ranges of 480-530 ppm are more likely than not and as likely as not to stay below 2°C. Emission levels for the year 2100 are indicated by the blocks to the right.

Source: IPCC (2014a)

Box 1.1. **The scientific basis** (*cont.*)

Emissions scenarios that are likely to keep global warming below 2°C (i.e. with at least a 66% probability) are characterised by GHG emissions reductions of between 40-70% by 2050 relative to 2010. This is a daunting challenge in comparison with the continued growth in GHG emissions – particularly CO_2 – over the last century or so (IPCC, 2014a). Global energy-related CO_2 emissions in 2012 were 50% higher than their 1990 level, at 31.6 billion tonnes of CO_2 ($GtCO_2$) (IEA, 2013a). According to the OECD *Environmental Outlook*, global GHG emissions could increase by another 50% by 2050 relative to 2010 unless strong abatement policies are implemented globally (OECD, 2012).

Cumulative emissions of CO_2 and other GHG will determine the extent of climate change (IPCC, 2014b).[2] Without mitigation efforts beyond those already in place, the IPCC projects increases in global mean surface temperature of between 3.7°C and 4.8°C by 2100 relative to the second half of the 19th century (IPCC, 2014a).[3] The scale and pace of this change would be unprecedented in human history. Regional changes at high latitudes and over the continents would be greater still. Precipitation patterns would be significantly affected both by an intensification of the hydrological cycle and changed atmospheric circulation patterns. The high latitudes and equatorial Pacific are projected to experience increased precipitation under high-emissions scenarios while mid-latitude and sub-tropical dry zones are likely to face reduced precipitation. Extreme precipitation is very likely to become more intense and more frequent over mid-latitude land masses and the wet tropics. These challenges will be exacerbated by continued melting of mountain glaciers and sea-level rise, which could reach as high as one meter by the end of the century in a high-emissions scenario, relative to 1986-2005 levels.

These changes would have strong impacts on the real economy. Concomitant effects on the financial sector can also be expected if valuable productive assets and infrastructures are affected on a large scale.

Note: 1. "So-called bio-energy with carbon capture and storage (BECCS) could be used in a wide range of applications, including biomass power plants, combined heat and power plants, flue gas streams from the pulp and paper industry, fermentation in ethanol production and biogas refining processes" (IEA, 2013a). 2. According to IEA (2014b) and IPCC (2014a) scenarios, the implementation of currently planned climate policies will put the world on a path that will have used up, by 2040, the upper estimate of the carbon budget available to keep the odds of staying below 2°C at 66%. This budget is estimated at 750-1 260 $GtCO_2$ for all greenhouse gases. 3. This range is for a median climate response; the full range is much larger.

Source: IPCC (2014a), *Climate Change 2014: Synthesis Report – Summary for Policymakers*, contribution of Working Groups I, II and III to the Fifth Assessment Report of the Intergovernmental Panel on Climate Change [core writing team, R.K. Pachauri and L.A. Meyer (eds.)], Intergovernmental Panel on Climate Change, Geneva, available at: www.ipcc.ch/pdf/assessment-report/ar5/syr/AR5_SYR_FINAL_SPM.pdf.

Climate change is part of a broader set of structural challenges faced today by policy makers: restoring robust economic growth, addressing systemic risk and instability in finance while ensuring investment in the real economy, slowing productivity, growing inequality in wealth distribution and persistent poverty, as well as ensuring the environmental sustainability of global economic development (OECD, 2015a).

These issues fall under different policy portfolios, but are interconnected. For instance, more long-term finance is needed for low-carbon and climate-resilient infrastructure; the decarbonisation of industrial products will require disruptive innovations, with possible implications for jobs; climate impacts will be mostly felt by vulnerable communities, particularly in developing countries. Reducing GHG emissions (especially CO_2 from fossil fuels) can foster a greener growth, provided that policies also address the potential distributive effects (OECD, 2013a).

For most countries, tackling climate change while pursuing greener, more inclusive growth is achievable (GCEC, 2014). Some investments in low-carbon infrastructure today can already provide attractive financial returns or energy savings which outweigh the additional upfront capital costs.[1] The transition to a low-carbon economy could also mean cleaner air, better health and a more diversified energy supply, in addition to the avoided impacts of warmer global climate.[2] In developing countries relying heavily on natural assets and particularly vulnerable to climate impacts, the low-carbon transition could also be a condition for long-term prosperity and well-being (OECD, 2013a).

Changing course to low carbon implies trade-offs in the short term, as some GHG-intensive activities would decline and others emerge, fossil fuels subsidies would need to be removed, and additional upfront capital would be needed to finance low-carbon assets. Understanding these changes will be essential to embark stakeholders on the necessary transformation of our economies to low carbon. International co-operation will be needed to help developing countries make the shift and manage short-term trade-offs associated with a transition to low carbon (see also Chapter 2 for a discussion on financing the transition).

The case for policy alignment

The policy direction to a low-carbon future has been clear for some time, even if the necessary concerted action has so far eluded the global community. Effective risk management demands strong policies to limit GHG emissions across activities and countries with due consideration to national circumstances. A three-pillar approach can be most effective in reducing GHG emissions: 1) a clear and robust price signal on GHG emissions or other market-based instruments; 2) smart regulations to remove market barriers to low-GHG choices; 3) forward-looking support to low-carbon technologies (see below). National circumstances, including the level of economic development and natural resource endowments, will affect the pace and breadth of implementation of these core climate policies and the potential short-term trade-offs, but this approach can create a framework to sustain growth over the long term.

The three-pillar approach is essential, and yet not in itself sufficient to effectively tackle climate change if our policy and regulatory systems continue to be geared towards CO_2-emitting fossil fuel technologies. Society's reliance on fossil fuels is responsible for the majority of global GHG emissions. Coal, oil and gas have fuelled most economic development for decades; they have influenced everything from the design of our homes, transport and cities to the way we produce food and manufacture goods. For instance, the use of natural gas to generate electricity in combined-cycle gas turbines influenced the liberalisation and design of electricity markets in some OECD countries in the 1990s. Similarly, today's urban planning and transport choices are marked by our reliance on petrol or diesel cars. The same observation applies to GHG-intensive inputs in agriculture and industry. Unravelling this will require a deep understanding of the policy signals driving it and an unwavering commitment to change.

This report presents the first broad diagnosis of the misalignments between overall policy and regulatory frameworks and climate goals. Left unchecked, they risk sending contradictory signals and hindering the low-carbon transition. Addressing these misalignments can enhance the responsiveness of economic and social systems to climate change policy efforts and facilitate the low-carbon transition. This, in turn, could help governments be more ambitious, both in domestic policies and in their international contributions.

Addressing climate policy misalignments forms part of the broader OECD Green Growth Strategy that seeks to avoid the recognised dangers of a return to business-as-usual after the financial and economic crisis (OECD, 2011a; 2015b). Climate policy has often not been at the very top of policy makers' priority lists globally, and yet this report shows that the implementation of sound climate policies and the removal of misalignments can positively contribute to some of policy makers' priority issues, including more resilient and inclusive growth.

Better alignment will often facilitate the achievement of other policy objectives and can make climate policy more acceptable to various stakeholders (e.g. a less polluted urban environment with more accessible city centres, tax code changes that reduce the most distortive taxes or more favourable conditions for long-term infrastructure investment in support of economic growth). Governments should identify and capitalise on these "win-win" outcomes.

Diagnosing misalignments

The multiple dimensions of misalignment

The analysis highlights policy misalignments across several dimensions:

- **Policy areas and policy objectives.** Is there consistency between goals, objectives or impacts of existing policy areas and climate policies? For instance, do financial market regulations have unintended negative consequences for low-carbon investments? Are tax systems encouraging CO_2-intensive development?
- **Development, economic and industrial policy goals.** Are policies that support development goals undermining long-term climate goals?
- **Levels of government.** Are the respective mandates of different levels of government and different ministries conducive to or hindering climate change objectives?
- **Stakeholders.** Do public and private actors have the same incentives for moving to low carbon – e.g. are potential climate risks transparently reflected in corporate disclosures and investor portfolios?
- **Borders.** Can one country's climate policy be undermined by another's domestic policy choices? Do international trade rules or unilateral trade remedies hinder the adoption of stronger climate policies? If so, how?

The policy experience used for this report is based on examples in OECD and, to a lesser extent on non-OECD countries. Countries present major differences related to their level of development, governance structure, institutional and financial capacity, natural resources and regulatory framework for the advancement of domestic climate policy. However, every country can take a comprehensive look across policy frameworks, make a diagnosis, and start addressing misalignments with climate goals for a more sustainable, low-carbon future.

Aligning policies across domains and activities

The report gives examples of misalignments of overall policy and regulatory frameworks with climate goals, and identifies a number of opportunities for realigning policies to enable an efficient and cost-effective shift to a low-carbon economy. The findings are presented in two parts:

- Cross-cutting policy domains which offer important channels for transformation: investment, taxation, innovation and skills, trade and adaptation.
- Specific activities that have a direct impact on emissions, and where climate objectives need to be systematically integrated into public and private decision-making processes: electricity, urban mobility and rural land use.

It is the sum of signals coming from misaligned policies which risks hindering the low-carbon transition, e.g. the combination of tax policies and urban planning rules encourage urban sprawl. Evaluating the effects of a constellation of policy signals on climate policy is ambitious, but the potential rewards in terms of efficiency and other gains are worth the effort.

From diagnosis to action

Aligning policies requires the engagement of parts of governments that have not so far been sufficiently mobilised in the development of climate response strategies, i.e. a better integration of climate policy with other structural policy challenges.

Climate policy can be made more effective if ministries with portfolios situated outside the traditional climate agenda can revisit the most misaligned policy instruments in their domains. A concerted whole-of-government effort is needed to establish a diagnosis, a mapping of climate policy instruments *and* of underlying policy frameworks that could help or hinder the transition to a low-carbon economy. The implementation phase of an ambitious and effective climate action plan, with core climate policies (a three-pillar approach described below) and the progressive resolution of other policy misalignments will also require new approaches to policy making across government (see Box 1.2).

Certain misalignments remain challenging, especially as their solutions do not hinge on national action alone. Concerns about distortions to competition, caused by climate policies with different levels of stringency across trade partners, continue to undermine climate policy ambition globally. This issue is limited to energy-intensive and internationally traded products such as aluminium, steel or chemicals. A global agreement could help in this area, although much will depend on the details of domestic policy implementation. Opportunities for more collaborative approaches or international co-ordination should be explored, for instance to guide innovation internationally rather than through country-specific measures.

Before turning to specific misalignments, the following section discusses the core package of climate policy that is needed for a transition to a low-carbon economy, and addresses some essential issues related to its acceptability.

Box 1.2. Government co-ordination: Insights from centres of government

A principal issue for governments with respect to aligning policies to promote the transition to a low-carbon economy is how co-ordinated policies can be implemented in practice given the complexity of the topic, the mixed track record of most governments in working horizontally, and the need to include an unprecedented range of public and private actors. The perspective of senior officials working at the centre of OECD governments, whose role is to provide strategic vision, policy co-ordination and monitoring for complex, cross-disciplinary policies, is that the low-carbon transition is indeed a unique challenge in terms of scale and time frame. As such, it requires new approaches to policy making across government. Governments have developed numerous solutions to establish more strategic co-ordination and better mainstreaming of climate policy objectives. These include super ministries, policy "tsars", inter-ministerial committees, independent policy units. On the policy front, options include legislations mandating national climate change targets (e.g. the United Kingdom's Climate Change Act) or impact assessments including guidance on how to include GHG emissions in these assessments. These can provide insights into the challenges and some of the solutions on which successful implementation will depend.

An ambitious effort to align policies requires several elements: a clear vision with measurable targets; an action plan with clear responsibilities and tasks for the different stakeholders; a system for monitoring progress; a process that has convening power, spans electoral cycles and engages opposition parties, and draws on co-ordination and substantive expertise.

To get to this degree of climate policy mainstreaming will require an investment in reflection on governance innovations best suited to this cross-portfolio issue. Overall, this requires the engagement of the head of government.

Source: Adapted from OECD (2014b), "Centre stage: Driving better policies from the Centre of Government", GOV/PGC/MPM(2014)3/FINAL, OECD, Paris, available at: www.oecd.org/gov/Centre-Stage-Report.pdf.

Core climate policies: Principles and implementation

Global GHG trends are *prima facie* evidence that existing climate policy instruments are inadequate in breadth (countries, sectors) and ambition (stringency). At the same time, there have never been more policy instruments implemented with climate mitigation as a primary objective or by-product – these provide a wealth of policy experience from which to draw lessons for more efficient GHG-reduction policy packages.

The variety of starting points and national circumstances precludes any kind of generalisation. The level and structure of a country's gross domestic product, its endowment in various resources, and physical and institutional infrastructure all affect the cost and feasibility of various policy options to curb GHG emissions. A country's openness to other economies and the perception of other countries' efforts to reduce emissions, i.e. the fairness of the global distribution of effort, are also critical factors in the willingness to take ambitious mitigation actions at home. The importance of a global agreement in Paris in 2015, under the aegis of the UNFCCC, should not be underestimated in this regard. In the end, however, the effectiveness of any accord hinges on the implementation of well-suited policies and measures at the level of national jurisdictions.

This section next recalls the basic principles of an appropriate policy package for an economically efficient mitigation strategy, before turning to important aspects of implementation, including the need to avoid misalignments or redundancy between policy tools, issues standing in the way of acceptability of these policies and possible solutions.

Basic principles for an efficient climate mitigation policy package

The gap between current and desired GHG emissions trends demonstrates the need for strong policies to reorient consumption, production and investment choices in our economies. The magnitude of the effort calls for close attention not just to immediate choices in the near term, but equally to the way policy settings are likely to influence the medium- and long-term costs of the transition to a low-carbon economy.

A three-pillar approach would provide a strong signal for the low-carbon transition:

- A robust price on GHG emissions with long-term credibility is a central pillar of any low-carbon economy, providing incentives for immediate emissions reductions where possible, as well as investment and innovation in low-GHG technologies. Market-based instruments and regulations deterring GHG-emitting activities also put a cost on emissions. Attempts to price carbon implicitly or explicitly will have distributional consequences that may be contentious. It requires the determination and creativity of governments to arbitrage between the economic efficiency and political and social sustainability of climate policies.
- Regulations may be particularly appropriate where a price signal is less effective due to market barriers or transaction costs – in particular in the household sector. These include emissions performance standards or measures to encourage energy efficiency.
- Targeted technology support (research, development and demonstration, or RD&D) can help to develop and lower the cost of risky but potentially promising sustainable low-GHG technologies, reducing the competitive gap with GHG-intensive technologies.

To be effective, these core climate policies must be backed by a clear long-term commitment by governments to support continuous and systematic efforts towards the transition to a low-GHG economy, giving private sector and civil society stakeholders the necessary confidence to take long-term decisions. A global agreement in Paris at COP21 in December 2015 would provide an essential impetus for countries to create such signals.

The challenge is to strike an appropriate balance between these three pillars, as well as to base them on efficient policy instruments. There are indeed major differences in the cost of policies mobilising various GHG mitigation potentials, as shown in *Effective Carbon Prices* (OECD, 2013b), sometimes a sign that policies have been inefficient.

Special consideration should also be given to possible interactions between the various policy instruments to ensure they are mutually reinforcing (Hood, 2011; 2013). In principle, GHG emissions pricing, the removal of market barriers (e.g. to support energy efficiency improvements in the household sector) and RD&D are complementary and can work to minimise the cost of GHG reductions better than a single instrument would (see, for instance, Acemoglu et al., 2012; Fischer and Newell, 2008).

The implementation of this policy package has not proven easy, however, even in countries that have pledged to reduce their GHG emissions. As described below, a few carbon pricing legislations have been rejected at various stages of legislative processes, and governments have had to move forward with other policy instruments. More generally, especially in developing countries, the question of energy affordability remains a barrier to ambitious climate action, even if practical solutions exist to mitigate these problems – e.g. a more efficient use of energy via energy efficiency measures, or cash transfers to reduce impacts on the poor.

The practice of climate mitigation policies

CO_2 pricing policies: A rapid overview

At present, carbon pricing policies only apply to a small portion of global GHG emissions – the list of regulatory measures that put an implicit price on CO_2 via a barrier to some uses of fossil fuels is probably far longer, but too diverse to be succinctly covered here.

Carbon taxes are in place in European countries such as Denmark, Finland, France, Iceland, Ireland, Norway, Slovenia, Sweden, Switzerland, Spain (with a tax on fluorinated gases, and carbon taxes in some of the country's autonomous communities); Costa Rica; Japan; Mexico; and in the Canadian provinces of Alberta and British Columbia; and a carbon tax is scheduled for introduction in Chile, Portugal and South Africa. As discussed in the chapter on taxation of this report (see Chapter 5), fossil fuels were taxed prior to the introduction of carbon taxes, and the distinction between carbon taxes and other excise taxes on the same energy products is blurred. The main policy features of carbon taxes are their coverage of a country's or region's CO_2 emissions, announced evolution, use of the collected revenues and treatment of distributional impacts. A later section addresses some of these from the viewpoint of political acceptability.

Emissions-trading systems stemmed initially from the Kyoto Protocol's flexibility mechanisms. The largest carbon market in operation is the EU Emissions Trading System, covering about 2 billion tonnes of CO_2 equivalent (tCO_2-eq) emissions. Several smaller systems have been implemented, with at least one initiative to link two systems (Quebec and California).[3] The systems vary in scope – e.g. few include forestry or agriculture activities, while almost all include power generation.[4] Under the auspices of the Kyoto Protocol, countries' binding emissions targets have also been used to stimulate emissions reductions in countries without emissions targets, via the Clean Development Mechanism (CDM). The CDM created a strong price incentive to reduce emissions in targeted activities in developing countries (including non-energy related activities, e.g. methane emissions from landfill), but diminished in importance when demand from EU Emissions Trading System (ETS) sources declined. Next in the evolution of emissions trading is the possibility of linking systems.[5] This would bring efficiency gains by broadening opportunities for emissions reductions at a lower cost to all participants – as indicated by existing price differences. The trade-off for domestic policy makers is loss of control over the policy-driven carbon price, and subsequent effects on other prices.

In addition to energy taxes, a number of OECD governments now systematically apply an explicit monetary carbon value in their appraisal of proposed policies and infrastructure decisions, e.g. based on a country's own assessment of the social cost of carbon (Smith et Braathen, 2015). As they conduct cost-benefit analyses of policies or projects, governments can factor in the impact on GHG emissions by applying this standard value (e.g. in USD/tCO_2-eq). This automatically advantages public policy choices that are less carbon-intensive even where there is no actual price paid for GHG emissions in the economy. These carbon values can be used in a number of instances, e.g. in a governmental agency's decisions on how to prioritise projects and policies; in governmental resource allocation plans based on social benefit criteria; in value-for-money assessment of public spending by auditing offices and parliamentary committees; or in independent studies of public policy choices (ibid.). Figure 1.2 shows carbon values used by some OECD countries for investment evaluations in the transport sector. The monetary values grow rapidly over time as their climate change impacts become increasingly severe.

Figure 1.2. **Monetary carbon values used for investment projects in the transport sector**

USD, 2014 value

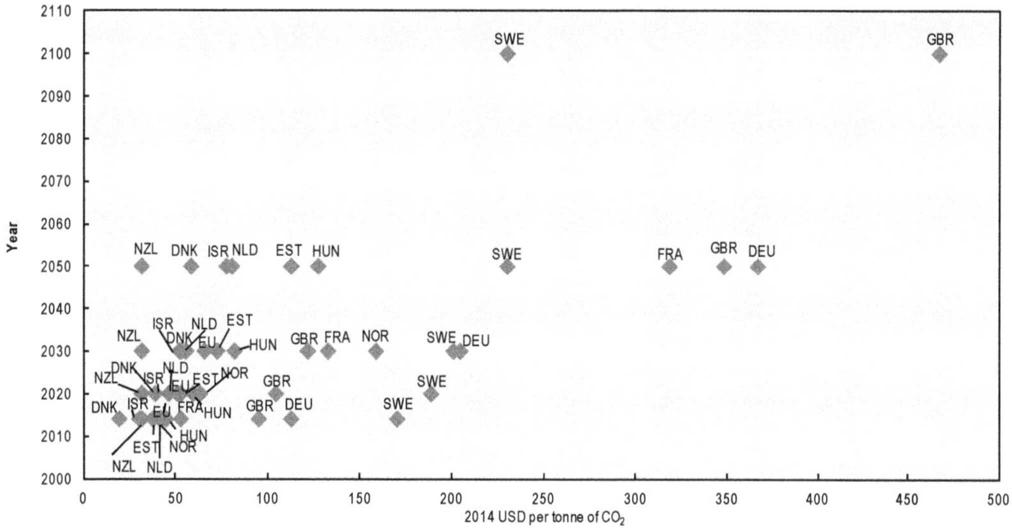

Source: Smith, S. and N.A. Braathen (2015), "Monetary carbon values in policy appraisal: An overview of current practice and key issues", *OECD Environment Working Papers*, OECD Publishing, Paris.

Avoiding misalignments between climate mitigation instruments

> In bridging the gap, governments should undertake an inventory of policies that explicitly and implicitly price carbon, and assess the impact of the policies and their interactions in order to ensure that they are mutually supportive in achieving CO_2 reductions as well as other social and economic objectives. (OECD, 2013c)

The vast majority of policy instruments introduced by countries to curb their GHG emissions are not within the carbon pricing pillar – recent examples include the United States' decision to regulate CO_2 emissions from power plants and possibly methane emissions from shale oil and gas operations and the People's Republic of China's (hereafter "China") decision has been made to replace all coal-based electricity generation with gas by the end of the decade. Table 1.1 gives an overview of the range of policy instruments (including pricing) in use to limit GHG emissions, primarily in the energy production and consumption sector.

Not all instruments affecting GHG emissions were introduced with emissions reductions as their primary goal and so should not be assessed against that metric alone. They nonetheless constitute the policy environment on top of which more ambitious climate policies will be introduced. The interactions between these instruments and those to come are an important element in the definition of an economically efficient mitigation strategy (Braathen, 2011; OECD, 2011b; Hood, 2011, 2013). These interactions in fact raise some alignment questions: will policies within the three pillars reinforce or undermine each other – i.e. will they reduce or add costs? Will interactions facilitate or hinder adoption by relevant economic agents? It is important to ensure that climate policies themselves are not in misalignment with one another before considering other policy areas.

Table 1.1. **A wide range of energy and climate policies to reduce greenhouse gas emissions**

Policy type	Policy options
Price-based instruments	– Taxes on CO_2 directly – Taxes/charges on inputs or outputs of process (e.g. fuel and vehicle taxes) – Subsidies for emissions-reducing activities – Emissions trading systems (cap and trade or baseline and credit)
Command and control regulations	– Technology standards (e.g. biofuel blend mandate, minimum energy performance standards) – Performance standards (e.g. fleet average CO_2 vehicle efficiency) – Prohibition or mandating of certain products or practices – Reporting requirements – Requirements for operating certification (e.g. Hydrofluorocarbon handling certification) – Land-use planning, zoning
Technology support policies	– Public and private research, development and demonstration (RD&D) funding – Public procurement – Green certificates (renewable portfolio standard or clean energy standard) – Feed-in tariffs – Public investment in underpinning infrastructure for new technologies – Policies to remove financial barriers to acquiring green technology (loans, revolving funds)
Information and voluntary approaches	– Rating and labelling programmes – Public information campaigns – Education and training – Product certification and labelling – Award schemes

Note: The bottom three categories are in the second and third pillars of climate mitigation instruments (market barrier removal and technology support).

Source: Hood (2011), "Summing up the parts: Combining policy instruments for least-cost climate mitigation strategies", International Energy Agency Information Paper, OECD/IEA, Paris, available at: www.iea.org/publications/freepublications/publication/Summing_Up.pdf, based on de Serres et al. (2010).

The combination of a tax on carbon emissions on the one hand and support to low-carbon supply technologies and energy efficiency on the other could, if well-designed, have one of two beneficial effects (Hood, 2013). It could either:

- increase emissions reductions beyond those achieved by the carbon tax, as the non-price measures could provide additional options that economic agents can seize to reduce carbon tax payments, or
- decrease the level of tax necessary to achieve a given level of emissions reductions.

In the case of an emissions-trading system in which the emissions limit is set, efficiency measures and support to low-carbon technologies should lower the market price of emissions allowances, as the cost of reduction is supported by other programmes. The question is then whether these other programmes impose a higher overall cost than what would be achieved with the emissions-trading system alone. The answer needs to take the following into account:

- If any of the measures in the other pillars tap into mitigation potentials that are lower than the market price, they lower the overall cost of meeting the emissions constraint.
- By lowering the market price of allowances such measures also lower its downstream effects, e.g. the increase in wholesale electricity prices via the pass-through of the allowances by power generators.

The OECD (2013b) identifies instances where the static cost of avoided CO_2 for certain support measures has been far greater than the observed CO_2 market price. While some subsidies to renewable technologies have been overly generous, these combined measures have eventually led to a considerable drop in the global cost of renewable technologies via learning and economies of scale. This is a clear advantage from a dynamic efficiency standpoint, as it lowers the price of CO_2 needed in the future to scale up deployment. While this dynamic efficiency argument cannot be an excuse for wasteful public policies, it should not be ignored in the economic assessment of a climate policy package.[6,7]

Policy interaction also occurs in the case of non-price policies. A regulation to close low-efficiency coal power plants would have cost repercussions for electricity users unless policies simultaneously encourage the uptake of electricity-saving appliances and behaviour. If not, more new electricity capacity would be necessary to replace coal, which could push electricity prices up. Interactions ought to be considered for all types of instruments to ensure that the overall package is in line with the overarching objective, whether it is a target level of emissions or the cost of avoided emissions.[8]

Enhancing the acceptability of ambitious climate mitigation

> In practice, governments have found it difficult to implement the most cost-effective and efficient policies for growth and reducing climate risk, such as carbon pricing. (GCEC, 2014)

If a number of countries have moved forward with more or less ambitious policy packages to curb their emissions, global GHG emissions trends are a reminder that policy lags behind the aspiration to limit global warming to 2°C, and that the introduction of instruments aiming at changing the direction of choices made in energy, industry and agriculture remains politically difficult. This section reviews some of the dimensions important for ensuring the acceptability of climate policies for economic activities and the electorate at large. Ensuring widespread buy-in to policies is essential if they are to be effective.

There are, of course, success stories showing how public opinion accepted, for instance, a tax on CO_2. Ireland introduced a carbon tax in the wake of the financial crisis (Convery et al., 2013), and British Columbia's carbon tax was successfully introduced and remained in place in spite of a lack of action by other North American jurisdictions (Harrison, 2013). But there are also examples of far-reaching, well-designed climate policies that have not managed to garner enough support to move forward with implementation. The failure of the Waxman-Markey proposal to introduce a GHG cap-and-trade system in the United States is a famous instance. Even where a legislature passes a measure, it can fail on constitutional grounds (e.g. the carbon tax legislation rejected by France's *Conseil Constitutionnel* or Constitutional Council in 2010).[9] Because climate policies inevitably run up against private interests, they will be the object of protests by various groups and lobbies. Governments, mindful of the need to maintain domestic economic growth and jobs, often significantly dependent on the rents from their own fossil fuel reserves, and sometimes under severe political pressure from financial backers, have in many cases pursued weaker climate policies.

From a pure economic perspective, climate policy is justified here and now to limit climate risks. The issue is how to cope adequately with the inevitable transition costs, because the transition to low carbon will create winners and losers. The energy and manufacturing sector will have to adopt new technologies and abandon others. Some activities may disappear as low-carbon products emerge and out-compete old ones. Similarly, consumers will face a new set of relative prices as policies steer them away from GHG-intensive choices. The distributional aspects of these changes, if left unchecked, can be a major barrier to reforms and have, in the past, hindered the adoption of ambitious climate policies.

Eliminating the policy alignment issues identified in the following chapters would lower the cost of the transition. Before delving into those broader misalignments, it is useful to dwell on three areas whereby core climate policies can be designed to consider transition costs: aligning climate efforts internationally, mitigating effects on industrial competitiveness and minimising distributional impacts for the poorer parts of society.

Building trust through international agreements

GHG emissions accumulating in the atmosphere have global effects, irrespective of where those emissions were released. This means that the climate impact of any one country's GHG reduction policies is dependent on action by others. Given that strong GHG mitigation policies may have short-term economic costs, the ambition of national policies is therefore dependent on the perceived impacts on national development that those policies may create, even if they will deliver longer term economic benefits. In a world of highly interconnected economies, governments – as well as industry and the general public – seek reassurance of commensurate action from their trade partners, through treaties or other forms of international agreement, whether bilaterally or multilaterally.

The UNFCCC remains the central international forum for global negotiations on climate change and also provides a window for co-ordination of climate policy. Such co-ordination can play an important role in the development of national GHG reduction policies in major economies, as it can provide assurance to domestic policy makers that commensurate efforts are being taken internationally by key trading partners. A global agreement under the UNFCCC would therefore be instrumental in providing political space for policy makers to strengthen climate action domestically. Indeed, all countries have been invited to make intended nationally determined contributions (INDCs) to global mitigation efforts before the Paris COP. A multilateral agreement may also help mobilise climate change finance for poor countries, both to help the most vulnerable adapt to climate change and to facilitate GHG mitigation.

Countries also pursue bilateral and regional agreements on climate change mitigation and adaptation in parallel to and in support of the UNFCCC negotiations. Decisions can be taken outside of the UNFCCC that align with and reinforce that process. An important example is the US-China joint announcement on climate change in November 2014 (White House, 2014). Such agreements are important to build trust between major economies. This can provide momentum to the UNFCCC negotiations as well as support domestic policy processes by demonstrating to stakeholders that climate action is being pursued internationally. Ideally, a virtuous circle of positive feedback would form between domestic policy progress and momentum in negotiations. In reality, domestic policy processes are rarely neatly aligned with the cycles of international negotiations.

The climate challenge is not alone in being a global problem requiring co-ordinated action from all major economies. In the modern globalised world of interconnected economies and multinational companies, capital is especially mobile. The OECD Base Erosion and Profit Shifting (BEPS) initiative has highlighted the extent to which OECD and other countries are losing revenue through tax evasion (OECD, 2013d).[10] Addressing BEPS requires universal action from major economies. If some countries act to close loopholes but others do not, capital and potential tax revenue will still flee to the latter. Over time, a consensus has built up that co-ordinated action is required. The OECD, in conjunction with the G20, is in the process of designing a multilateral agreement and accompanying rules to tackle BEPS internationally. This is a good example of government consensus achieving powerful objectives despite strong financial interests pulling in the other direction. In some ways, a similar consensus exists for action on climate change mitigation. However, in many countries strong opposition to more stringent climate policies has been voiced by energy-intensive industries.

The impact of CO_2 mitigation on various parts of the economy

Industrial competitiveness: A thorn in the side of climate ambition

An important aspect of the political acceptability of ambitious climate mitigation policy is how easily industry can adjust and thrive under this new constraint. The question is highly relevant for industries with a high-carbon or high-energy content.

Energy-intensive firms in many countries remain concerned that if domestic climate-related regulation is misaligned with the stringency of regulation in other countries, this will harm competitiveness at the firm and sector level and could lead to industrial flight to countries with less stringent climate regulation, with corresponding economic and employment impacts in the original country. Emissions reduction efforts would also be undermined, as part of the avoided emissions would now occur somewhere else. This potential "carbon leakage" to "pollution havens" has been much discussed in the literature (see examples in Condon and Ignaciuk, 2013; and Arlinghaus, 2015).

Various recent OECD studies have explored the relationship between environmental policies, in particular climate policies, and economic performance. A detailed analysis of environmental policy and productivity also shows that such policies have not, so far, been detrimental for overall growth. On the contrary, stringent environmental policies can stimulate productivity growth in the short term for technologically advanced countries and firms (Albrizio et al., 2014). Separately, other OECD work (Sauvage, 2014) found that greater environmental stringency can lead to increased exports of some environmental goods.

The literature on competitiveness impacts of climate policy is based on two approaches whose conclusions differ widely on the seriousness of competitive distortions in climate policy:

- empirical studies taking an econometric look at the industry response to climate policy to date
- modelling simulations that project future climate policy in multi-regional frameworks over the medium to long term.

A review of empirical studies found very little evidence of sector-level competitiveness effects arising from carbon pricing systems implemented to date (Arlinghaus, 2015). While the literature is in broad agreement that the EU ETS has stimulated some emissions abatement, no causal link could be established between carbon pricing – including the EU ETS and a range of carbon taxes – and carbon leakage. For carbon taxes, while abatement through decreases in energy intensity was found, only very small impacts on competitiveness were identified (ibid.). Further, no causal effects of the system on employment, output or international trade have been found; observed employment decreases are more likely due to the financial crisis and the decades-long gradual shift away from manufacturing in OECD countries (Warwick, 2013; Pilat et al., 2006). The review also found that pass-through of carbon costs to consumers is sometimes higher than expected, even in industries exposed to international trade (Arlinghaus, 2015). The main question looking at empirical evidence is whether conclusions would apply for emissions reduction levels that are more significant than those observed to date. GHG-intensive industries consistently argue that their potential for significant emission reductions are limited in the near to medium term unless they outsource their most GHG-intensive processes to other regions, leading to carbon leakage.

Macroeconomic modelling studies simulate positive leakage rates, generally below 20%, meaning that one-fifth of emissions reductions achieved in a region would in fact be offset by higher emissions in another region. This would come as its industry becomes more competitive, and also as a result of lower international energy prices leading to higher energy consumption in regions without a greenhouse gas constraint (IPCC, 2014b; Condon and Ignaciuk, 2013). These rates are computed on the basis of emissions reductions achieved by whole economies, i.e. not focusing on industry-by-industry leakage, which could be much higher.

Almost all carbon pricing schemes have included measures to lessen the impacts on energy-intensive, trade-exposed sectors. These measures include full or partial free tradable permits, rebates on taxes or other financial measures. In theory, governments could also implement border-based protection measures such as carbon-based tariffs or border carbon adjustments, but none have yet done so in practice (Condon and Ignaciuk, 2013). While a potentially effective tool for reducing competitiveness effects of strong climate policy, such measures would be fraught with political and technical difficulties. On the political side, they have a clear impact on international trade and so could be challenged through the World Trade Organization if they are interpreted as being more about protectionism than climate change. On the technical side, it would be challenging to accurately assess product-specific emissions of goods arriving at the border. In the case of an emissions-trading system with a fluctuating carbon price, setting the precise tariff level would not be easy (WTO, 2013).

The debate on industrial competitiveness and climate policy remains open and sensitive as regions undertake mitigation at different paces. The absence of evidence of competitiveness effects to date can be challenged on the ground that future emissions reduction levels will need to be much higher than implemented so far, with higher costs and possible trade distortions as a result. This, of course, hinges on the ambition of future climate policies in countries that have not yet implemented GHG constraints on their industry. Further, the industrial competitiveness landscape will also be influenced by the evolution of domestic energy prices (IEA, 2013b; Flues and Lutz, 2015). The cost of climate policy is one of many factors in this picture; energy costs, labour costs, exchange rates, transport costs, product specialisation, and local demand markets and regulations are important determinants of industrial competitiveness (IEA, 2013b; ECF, 2014). A global UNFCCC agreement on GHG mitigation could help, although much eventually depends on details of policy implementation across regions.

Opportunities for more collaborative approaches should be explored, for instance the possibility of pooling innovation efforts internationally on industrial breakthrough technologies for the low-carbon transition of these large emitters. How domestic policies will affect them and what remedies may be applied is also an area where policy makers could gain from sharing best practice.

How will household incomes be affected by the low-carbon transition?

Although the taxation of carbon may not be part of the policy toolkit for all countries in the near term, there is empirical experience on how to cope with the distributional effects of such taxes. Modelling studies add to this empirical knowledge, especially on how to best recycle revenues from a new tax on carbon, either to improve economic efficiency or to mitigate negative effects of the tax on social equity. Tax revenue recycling is only an option if climate policy instruments raise tax revenues, which is not the case of subsidies and regulatory approaches (Heindl and Löschel, 2014). This reinforces the case for pricing carbon through taxation or auctioned emissions allowances.

How climate mitigation policy will affect various segments of the population depends on a range of factors, including income distribution, the level of emissions of various household-related activities (e.g. is electricity low- or high-carbon?) and relative prices, but also on the kind of policy instruments implemented to reduce emissions. Sutherland (2006) illustrates, for instance, how an energy efficiency standard for appliances can have a regressive effect, even if less visible than a fuel price increase from a tax.

A recent and comprehensive survey of modelling studies of carbon taxes shows these instruments to be generally regressive, but the conclusion should not stop here.

> The review of literature on distributional effects of energy and carbon taxation reveals two key findings. Firstly, direct energy and carbon taxation is regressive in many cases. This implies that low-income households are responsible for proportionally larger tax burdens compared to wealthier households. Secondly, such negative effects usually can be fully neutralised or even reversed if appropriate changes are made to existing tax and benefit schemes in parallel to the new energy or carbon tax. In this case, poorer households could even benefit from the reform. (Heindl and Löschel, 2014)

The overall effect on income for different segments of society results from at least two mechanisms pulling in different directions (Flues, 2015; Heindl and Löschel, 2014):

- Some energy goods such as electricity, and gas for cooking and heating, are often a necessity, and their share in household income and expenditures tends to be higher for poorer segments of society. A climate policy with a high impact on electricity prices would tend to be regressive.
- Car fuel usage tends to increase with income in OECD countries. At present, taxes on fuels for car use are not on average regressive when observed for 21 OECD countries, mostly European. An increase in fuel taxes would rather have a progressive effect. As an extreme illustration, poor households that do not own a car would not be directly exposed to an increase in fuel taxes.

Another dimension of the distributive effects of higher energy prices and taxes is the geographical situation of targeted households. Generally, studies find that taxes fall more on rural than urban households, as the former rely more on personal vehicles and drive longer distances; they are also more likely to live in detached houses with higher heating fuel consumption (Flues, 2015). It is not clear, however, that this warrants special treatment. If distributional issues are to be addressed, they should eventually be based on effects on overall income, irrespective of other factors. The question is whether short-term measures would need to be implemented to facilitate the adjustment of those more exposed to price changes.

A number of options to recycle energy and carbon tax revenues have been studied and implemented, including: income-tested benefits, lump-sum transfers (equal amount per household or individual); lower social security or other labour charges; a reduction of value-added tax (VAT); or investment tax credits, with varying effectiveness with respect to distributional impacts. As indicated above, governments can offset and even reverse the regressive nature of higher carbon and energy taxes: an income-tested benefit financed by additional tax revenues can target lower income households, making them better off with higher energy taxes than without. A lump-sum transfer (i.e. equal amount per household) would also disproportionally benefit lower income households.

In summary, governments must balance the following considerations in order to address the distributional issues of the low-carbon transition, as considered through the prism of a carbon price on energy:

- The regressive or progressive nature of the increase in energy prices and other regulatory interventions.
- Whether and how to enhance the efficiency of the tax system: the reduction of other taxes can have beneficial impacts on the economy through effects on the productive sector, lower labour costs and less distortive taxes generally.
- Whether the regressive effects of the tax ought to be addressed through specific tax expenditures, cash transfers and other social benefits.

The main result from analyses to date is that expenditures to offset the negative impacts on the less well-off are usually lower than the revenues raised by a new carbon tax (Heindl and Löschel, 2014). In other words, this particular barrier to the acceptability of climate mitigation policy can be overcome.

Properly accounting for co-benefits

The acceptability of low-carbon policies also depends on the set of economic, environmental and social co-benefits brought by various GHG policy instruments. The IEA (2014a) recently explored the multiple benefits of energy efficiency measures, beyond energy savings at end-use level. These benefits range from health to industrial productivity, to broader macroeconomic effects and climate resilience via a lesser need for potentially vulnerable energy infrastructure. Capturing the value of all the associated benefits is difficult, but assigning them a zero value is hardly justifiable. As a striking counter-example, the OECD (2014a) recently quantified the cost of air pollution from NO_x and particulate matter from diesel use in transport to some USD 850 billion annually in OECD countries, based on estimated value of lives lost and of ill health; fuel-savings in transport would therefore have both CO_2 and health benefits.

GHG mitigation policies will also have indirect effects that may lower their intended effectiveness. Since the first oil shock and the introduction of policies to save energy, energy policy analysts have been debating the importance of the rebound effect: the expected energy savings would be partly offset by some additional energy use enabled by the primary savings. A very comprehensive study conducted by the UK Energy Research Centre finds that “the evidence does not suggest that improvements in energy efficiency routinely lead to economy-wide increases in energy consumption” (Sorrell, 2007). In the context of a three-pillar approach to climate policy (carbon pricing, energy efficiency policy, RD&D support), the rebound effect would be further minimised by an increase in energy costs.

Box 1.3. Trade-offs between local pollution and climate change policy goals

The fight against rising GHG emissions can go hand-in-hand with the resolution of local environmental problems that have very direct impacts on local populations. Policy intervention in such cases is facilitated by support for policy measures that abate local pollution and GHG emissions at the same time.

However, this is not always the case. Some solutions to local environmental problems can, in fact, contribute to higher greenhouse gas emissions: e.g. traditional equipment to remove SO_x and NO_x from coal-based power generation lowers a plant's thermal efficiency, and so raises CO_2 intensity (the reduction of aerosols to increase the greenhouse effect). It is, however, possible to combine high-efficiency coal plants with low-NO_x burners.

Another example is the simultaneous development of coal-based synthetic natural gas (SNG) in China and its possible use in city centres to replace coal in order to reduce air pollution. Recent research shows the impact that this strategy would have on the country's CO_2 emissions. The carbon footprint of SNG is estimated to be seven times that of natural gas, and current plans do not currently include carbon capture and storage, which could significantly reduce emissions from SNG production. The use of SNG in electricity would release 36-82% more CO_2 than coal. "If all 40 or so of the projected facilities are built, the GHG emissions would be an astonishing ~110 billion tonnes of CO_2 over 40 years" (Yang and Jackson, 2013). The country's plan to introduce a nationwide emissions trading system in the coming years is likely to contradict such plans. They are nonetheless a vivid example of how a solution to local environmental problems does not automatically lead to lower GHG emissions (see IEA, 2014d for a discussion of pollution control in China and expected climate benefits).

With possible interactions and multiple costs and benefits, a systematic methodology is required to integrate all available information to guide decision making. A thorough multi-dimensional cost-benefit analysis (CBA), i.e. a regulatory impact assessment, is a useful starting point. One must recognise that the value used for GHG emissions in such a methodology will be imprecise given the structural uncertainties related to climate change. Nevertheless, however imprecise a monetary metric may be to capture dimensions as diverse as health and environmental impacts, it is far superior to an implicit valuation at zero of all non-financial dimensions (see OECD, 2006 for a discussion of CBA for environmental policy). A very useful contribution of the CBA process is the identification of potential winners and losers from the proposed policy; such understanding is essential for making a proper case for the introduction of a policy change.

Notes

1. On a global scale between 2011 and 2050, the International Energy Agency's *Energy Technology Perspectives*' 2°C Scenario (2DS) generates fuel savings that outweigh the additional investment required to keep CO_2 emissions from energy on a 2°C-compatible path, including when applying a 10% discount rate (IEA, 2014a). See also IEA (2014b).
2. See OECD (2014a) for a discussion of air quality effects of fossil fuel use in transport.
3. Other systems in place include national-level systems in New Zealand and Korea, North America's Regional Greenhouse Gas Initiative, China's provincial and cities carbon market pilots (which should evolve into a nationwide system by 2020), Japan's Tokyo and Saitama prefectures' systems, and Kazakhstan's pilot ETS (Alberola, 2014; World Bank, 2014a).
4. The World Bank Partnership for Market Readiness provides a forum for national and regional policy makers to exchange on the design and implementation details of carbon market instruments (World Bank, 2014b).
5. Norway has already linked its system with the EU ETS and Switzerland may follow soon.
6. Recent theoretical work supports that an optimal solution to long-term CO_2 mitigation should combine a carbon price and R&D expenditures to reduce the cost of not-yet-competitive CO_2-saving technologies. Without directed support it would take a much higher carbon price to encourage such R&D, with a higher welfare cost overall (Acemoglu et al., 2012).
7. Another dynamic effect is mentioned in OECD (2011b): policies that inadvertently result in additional emission reductions – i.e. a lower price of CO_2 on the market can also facilitate the adoption of a more ambitious emission reduction if and when future emission caps are set by the regulator.
8. For detailed recommendations on this issue, see Hood (2011; 2013) and OECD (2013b).
9. Another carbon tax was introduced in France starting in 2014.
10. BEPS refers to tax planning strategies that exploit gaps and mismatches in tax rules to artificially shift profits to low or no-tax locations where there is little or no economic activity, resulting in little or no overall corporate tax being paid. BEPS is of major significance for developing countries due to their heavy reliance on corporate income tax, particularly from multinational enterprises (OECD, 2013a).

References

Acemoglu, D. et al. (2012), "The environment and directed technical change", *American Economic Review*, American Economic Association, Vol. 102, No. 1, February, http://dx.doi.org/10.1257/aer.102.1.131.

Alberola, E. (2014), "EU ETS et autres ETS émergents: Perspectives de coopération", presentation to the 35th session of the Conseil Économique du Développement Durable, 19 November 2014, ministère de l'Écologie, du Développement durable et de l'Énergie, Paris.

Albrizio, S. et al. (2014), "Do environmental policies matter for productivity growth? Insights from new cross-country measures of environmental policies", *OECD Economics Department Working Papers*, No. 1176, OECD Publishing, Paris, http://dx.doi.org/10.1787/5jxrjncjrcxp-en.

Arlinghaus, J. (2015), "Competitiveness impacts of carbon pricing: A review of empirical findings", *OECD Environment Working Papers*, No. 87, OECD Publishing, Paris, http://dx.doi.org/10.1787/5js37p21grzq-en.

Braathen, N.A. (2011), "Carbon-related taxation in OECD countries and interactions between policy instruments", Chapter 1 in: Kreiser, L. et al. (eds.), *Environmental Taxation and Climate Change: Achieving Environmental Sustainability through Fiscal Policy*, Critical Issues in Environmental Taxation series, Volume X, Edward Elgar.

Condon, M. and A. Ignaciuk (2013), "Border carbon adjustment and international trade", *OECD Trade and Environment Working Papers*, No. 2013/06, OECD Publishing, Paris, http://dx.doi.org/10.1787/5k3xn25b386c-en.

Convery, F.J. et al. (2013), "Ireland's carbon tax and the fiscal crisis: Issues in fiscal adjustment, environmental effectiveness, competitiveness, leakage and equity implications", *OECD Environment Working Papers*, No. 59, OECD Publishing, Paris, http://dx.doi.org/10.1787/5k3z11j3w0bw-en.

de Serres, A. et al. (2010), "A framework for assessing green growth policies", *OECD Economics Department Working Papers*, No. 774, OECD Publishing, Paris, http://dx.doi.org/10.1787/5kmfj2xvcmkf-en.

ECF (2014), "Europe's low-carbon transition: Understanding the challenges and opportunities for the chemical sector", European Climate Foundation, Brussels, March, available at: http://europeanclimate.org/wp-content/uploads/2014/03/ECF-Europes-low-carbon-Transition-web1.pdf.

Fischer, C. and R. Newell (2008), "Environmental and technological policies for climate mitigation", *Resources for the Future Discussion Paper*, RFF DP 04-05, April, revised February 2007, Resources for the Future, Washington, DC, available at: www.rff.org/rff/Documents/RFF-DP-04-05-REV.pdf.

Flues, F. (2015), "The distributional effects of energy taxes", *OECD Environment Working Papers*, OECD Publishing, Paris, forthcoming.

Flues, F. and B. Lutz (2015), "Competitiveness effects of the German electricity tax", *OECD Environment Working Papers*, OECD Publishing, Paris, forthcoming.

GCEC (2014), *Better Growth, Better Climate: The New Climate Economy Report*, The Global Commission on the Economy and Climate, Washington, DC, October, available at: http://newclimateeconomy.report/wp-content/uploads/2014/08/GCEC_GlobalReport.pdf.

Harrison, K. (2013), "The political economy of British Columbia's carbon tax", *OECD Environment Working Papers*, No. 63, OECD Publishing, Paris, http://dx.doi.org/10.1787/5k3z04gkkhkg-en.

Heindl, P. and A. Löschel (2014), "Energy sector reform and its impact on households", Issue Note, OECD Green Growth and Sustainable Development Forum, 13-14 November 2014, OECD, Paris, available at: www.oecd.org/greengrowth/Issue%20Note%20Session%20One%20GGSD%20Forum.pdf.

Hood, C. (2013), "Managing interactions between carbon pricing and existing energy policies: Guidance for policymakers", International Energy Agency Insights Series 2013, OECD/IEA, Paris, available at: www.iea.org/publications/insights/insightpublications/managinginteractionscarbonpricing_final.pdf.

Hood, C. (2011), "Summing up the parts: Combining policy instruments for least-cost climate mitigation strategies", International Energy Agency Information Paper, OECD/IEA, Paris, available at: www.iea.org/publications/freepublications/publication/Summing_Up.pdf.

IEA (2014a), *Capturing the Multiple Benefits of Energy Efficiency: A Guide to Quantifying the Value Added*, International Energy Agency, Paris, http://dx.doi.org/10.1787/9789264220720-en.

IEA (2014b), *Energy, Climate Change and Environment, 2014 Insights*, International Energy Agency, Paris, http://dx.doi.org/10.1787/9789264220744-en.

IEA (2013a), *Redrawing the Energy-Climate Map: World Energy Outlook Special Report*, International Energy Agency, Paris, available at: www.iea.org/publications/freepublications/publication/WEO_RedrawingEnergyClimateMap.pdf.

IEA (2013b), *World Energy Outlook 2013*, International Energy Agency, Paris, http://dx.doi.org/10.1787/weo-2013-en.

IPCC (2014a), *Climate Change 2014: Synthesis Report – Summary for Policymakers*, contribution of Working Groups I, II and III to the Fifth Assessment Report of the Intergovernmental Panel on Climate Change [core writing team, R.K. Pachauri and L.A. Meyer (eds.)], Intergovernmental Panel on Climate Change, Geneva, available at: www.ipcc.ch/pdf/assessment-report/ar5/syr/AR5_SYR_FINAL_SPM.pdf.

IPCC (2014b), *Climate Change 2014: Mitigation of Climate Change*, contribution of Working Group III to the Fifth Assessment Report of the Intergovernmental Panel on Climate Change, Edenhofer, O. et al. (eds.), Cambridge University Press, Cambridge, United Kingdom and New York, New York, United States.

OECD (2015a), *New Approaches to Economic Challenges – Final Synthesis Report*, OECD, Paris.

OECD (2015b), *Towards Green Growth? Tracking Progress*, OECD Publishing, Paris, http://dx.doi.org/10.1787/9789264234437-en

OECD (2014a), *The Cost of Air Pollution: Health Impacts of Road Transport*, OECD Publishing, Paris, http://dx.doi.org/10.1787/9789264210448-en.

OECD (2014b), "Centre stage: Driving better policies from the Centre of Government", GOV/PGC/MPM(2014)3/FINAL, OECD, Paris, available at: www.oecd.org/gov/Centre-Stage-Report.pdf.

OECD (2013a), *Putting Green Growth at the Heart of Development*, OECD Green Growth Studies, OECD Publishing, Paris, http://dx.doi.org/10.1787/9789264181144-en.

OECD (2013b), *Effective Carbon Prices*, OECD Publishing, Paris, http://dx.doi.org/10.1787/9789264196964-en.

OECD (2013c), "Climate and carbon: Aligning prices and policies", *OECD Environment Policy Papers*, No. 1, OECD Publishing, Paris, http://dx.doi.org/10.1787/5k3z11hjg6r7-en.

OECD (2013d), *Action Plan on Base Erosion and Profit Shifting*, OECD Publishing, Paris, http://dx.doi.org/10.1787/9789264202719-en.

OECD (2012), *OECD Environmental Outlook to 2050: The Consequences of Inaction*, OECD Publishing, Paris, http://dx.doi.org/10.1787/9789264122246-en.

OECD (2011a), *Towards Green Growth*, OECD Green Growth Studies, OECD Publishing, Paris, http://dx.doi.org/10.1787/9789264111318-en.

OECD (2011b), "Interactions between emission trading systems and other overlapping policy instruments", *OECD Green Growth Papers*, No. 2011/02, OECD Publishing, Paris, http://dx.doi.org/10.1787/5k97gk44c6vf-en.

OECD (2010), *Tax Policy Reform and Economic Growth*, OECD Tax Policy Studies, No. 20, OECD Publishing, Paris, http://dx.doi.org/10.1787/9789264091085-en.

OECD (2006), *Cost-Benefit Analysis and the Environment: Recent Developments*, OECD Publishing, Paris, http://dx.doi.org/10.1787/9789264010055-en.

Pilat, D. et al. (2006), "The changing nature of manufacturing in OECD economies", *Science, Technology and Industry Working Papers*, No. 2006/9, OECD Publishing, Paris, http://dx.doi.org/10.1787/308452426871.

Sauvage, J. (2014), "The stringency of environmental regulations and trade in environmental goods", *OECD Trade and Environment Working Papers*, No. 2014/03, OECD Publishing, Paris, http://dx.doi.org/10.1787/5jxrjn7xsnmq-en.

Smith, S. and N. A. Braathen (2015), "Monetary carbon values in policy appraisal: An overview of current practice and key issues", *OECD Environment Working Papers*, OECD Publishing, Paris.

Sorrell, S. (2007), *The Rebound Effect: An Assessment of the Evidence for Economy-wide Energy Savings from Improved Energy Efficiency*, report produced by the Sussex Energy Group for the Technology and Policy Assessment function of the UK Energy Research Centre, October, available at: www.ukerc.ac.uk/asset/3B43125E-EEBD-4AB3-B06EA914C30F7B3E.

Sutherland, R.J. (2006), "The distributional effects of direct regulation: A case study of energy efficiency appliance standards", in: Johnstone, N. and Y. Serret (eds.) (2006), *The Distributional Effects of Environmental Policy*, Edward Elgar Publishing.

Warwick, K. (2013), "Beyond industrial policy: Emerging issues and new trends", *OECD Science, Technology and Industry Policy Papers*, No. 2, OECD Publishing, Paris, http://dx.doi.org/10.1787/5k4869clw0xp-en.

White House (2014), *U.S.-China Joint Announcement on Climate Change*, The White House website, Office of the Press Secretary, 11 November 2014, www.whitehouse.gov/the-press-office/2014/11/11/us-china-joint-announcement-climate-change (accessed 2 April 2015).

World Bank (2014a), *State and Trends of Carbon Pricing 2014,* The World Bank Group, Washington, DC, May, http://dx.doi.org/10.1596/978-1-4648-0268-3.

World Bank (2014b), "The World Bank Partnership for Market Readiness", The World Bank Group, Washington, DC, www.thepmr.org.

WTO (2013), *World Trade Report 2013: Factors Shaping the Future of World Trade*, World Trade Organization, Geneva, available at: www.wto.org/english/res_e/booksp_e/world_trade_report13_e.pdf.

Yang, C. and R.B. Jackson (2013), "China's synthetic natural gas revolution", Commentary, *Nature Climate Change*, Vol. 3, pp. 852-854, October, Nature Publishing Group, http://dx.doi.org/10.1038/nclimate1988.

Part I

Reforming cross-cutting policies

Chapter 2

Scaling-up low-carbon investment and finance

There is an urgent need to significantly scale-up investment to low-carbon, more energy efficient alternatives (e.g. renewable energy, sustainable transport systems, energy efficiency) and to shift investment away from fossil fuel use. The low-carbon transition will require mobilising of all sources of public and private sector investment and finance, including institutional investors. Governments need to use their scarce resources to trigger large-scale private sector investment in activities otherwise unlikely to attract sufficient private funding. However, some financial systems regulations hinder the allocation of long-term finance to low-carbon infrastructure investments. This chapter reviews such barriers in areas including: regulations related to long-term investment; corporate disclosures on climate risks; public procurement; and the allocation and delivery of development finance. It then provides some guidance for governments on how to align principles governing financial regulations, corporate governance and public spending with the low carbon transition.

Key messages

The global economy requires around USD 90 trillion of investment in infrastructure between 2015 and 2030 to support economic development (e.g. buildings, transport, energy). Investing in low-carbon, climate-resilient infrastructure could put the world on a 2°C trajectory and deliver significant co-benefits, including improvements in air quality, health, energy savings and better mobility. In contrast, decisions taken today on infrastructure such as transport, power generation and buildings could lock economies deeper into greenhouse gas-intensive systems, technologies and innovation and make them vulnerable to a changing climate.

The low-carbon transition investment challenge is twofold, involving:

1. *Scaling-up finance* for long-term investment in infrastructure.
2. *Shifting investments* towards low-carbon alternatives.

Investment is not moving significantly away from carbon-intensive technologies and infrastructure. Current market forces and regulations tend to collectively favour investment in fossil fuel activities over low-carbon infrastructure, often unintentionally. Governments need to better understand the regulatory, legal and governance impediments to the supply of long-term finance, and act rapidly to align market expectations with a low-carbon future.

The following questions highlight some core misalignments:

Is the current investment framework aligned with the low-carbon transition? Strong, stable climate policies are necessary to adjust the return on investment of low-carbon infrastructure projects, but on their own are not enough. Policy makers need to address a range of policy misalignments in the overall investment framework that collectively favour investment in fossil fuel-intensive activities. These include conflicting incentives in competition, trade, tax and innovation policies as well as insufficient institutional settings to enhance co-ordination between levels of government (explored further in other chapters of this report). Promoting competitive neutrality and keeping markets open will be key to maximising foreign and domestic investment, enhancing competitive pressure, promoting innovation and reducing costs.

Is the regulatory framework for investors and financiers conducive to low-carbon, long-term investments? Regulatory frameworks implemented to pursue objectives outside the climate sphere can have unintended consequences. For instance, the financial crisis has motivated changes to financial stability rules – e.g. Basel III – that are sometimes accused of inadvertently limiting the ability of institutions such as banks to finance long-term infrastructure investments. While financial stability is a prerequisite to any kind of investment, including low-carbon investment, an appraisal of the potential unintended impact of rules governing the financial sector (accounting, prudential, market) on the supply of long-term finance appears necessary. In particular, regulatory-induced misalignments between the time horizons of investors and the need for long-term infrastructure financing would undermine the low-carbon transition. Greater transparency and harmonisation of corporate disclosures on climate risks and liabilities could also encourage climate-friendly investments.

Are climate goals mainstreamed in public spending and development policies? Public finance and investment could be powerful catalysts for the low-carbon transition, yet they are not fully aligned with climate goals. Although its share has been rising over time, less than one-fifth of official development assistance (ODA) is climate-related. Public support to private investments in coal and gas is still significant. National and international public financial institutions could more systematically lead the way by reconsidering their support to greenhouse gas-intensive projects.

The low-carbon investment challenge

Unprecedented economic, social and technological transformation is needed to limit global warming to 2°C above industrial levels and avoid catastrophic climate change. This will require a dramatic reallocation of investment away from carbon-intensive infrastructure to sustainable[1] low-carbon and climate-resilient infrastructure, such as solar photovoltaic arrays and wind farms, energy efficient buildings, smart grids, public transport and electric vehicle charging stations (IEA, 2014). Investment policies need to create the conditions for this reallocation.

Irrespective of climate change concerns, the global economy requires around USD 90 trillion of investment in infrastructure (e.g. buildings, transport, energy) between 2015 and 2030 to support economic growth and the broader development agenda (GCEC, 2014). In advanced economies, many ageing infrastructure networks for water, energy and transport need to be replaced or upgraded. In emerging and developing economies, most of the infrastructure required to meet development goals is still to be built, particularly in urban settings (Corfee-Morlot et al., 2012). Historically, the challenge of rising energy demand and transport needs has mostly been met with fossil fuels. City infrastructure mainly revolves around the use of private cars.

Today there is an unprecedented opportunity to ensure that new investment in infrastructure helps to mitigate and adapt to climate change while also ensuring economic development, energy security and safe and reliable transport for all. While some investments in carbon-efficient fossil-fuel infrastructure is still needed to support economic development in the coming decades, there is an urgent need to reallocate and significantly scale up capital to low-carbon, more energy efficient alternatives. Failing to seize this opportunity will lock in CO_2 emissions for decades to come, implying significantly higher social and economic costs of mitigation action (OECD, 2012a).

The incremental short-term costs of shifting to low carbon would amount to just a fraction of the finance needed for infrastructure overall (Box 2.1). These additional costs are limited compared to the potential impacts of climate change if this action is not taken (Dellink et al., 2014). In addition, low-carbon infrastructure would bring many other benefits: better mass transport in cities can reduce congestion and air pollution; distributed renewable energy infrastructure projects can improve access to energy in developing countries while lowering emissions, improving indoor air quality and stimulating innovation through technology transfer and international co-operation (OECD, 2011; 2012a; 2015a); and making buildings more energy efficient could reduce energy costs. These long-term costs and benefits should be fully taken into account.

Closing the financing gap

Irrespective of climate change, many countries are struggling to mobilise long-term finance to meet infrastructure needs of all kinds. According to the World Economic Forum, global spending on basic infrastructure – transport, power, water and communications – amounts to USD 2.7 trillion a year when it ought to be USD 3.7 trillion (WEF, 2014).

Box 2.1. Investment needs for a low-carbon energy sector under the IEA 450 Scenario

The International Energy Agency (IEA) estimates that cumulative investment of USD 53 trillion in energy supply and efficiency will be needed by 2035 to achieve the goal of keeping global warming below 2°C. This is only about 10% more than the USD 48 trillion that would be needed in the sector irrespective of climate change.

The composition of this investment shifts away from fossil fuels (USD 4.3 trillion lower) and towards the power sector, particularly renewables, carbon capture and storage (CCS) technologies and nuclear. Investments in energy efficiency are higher by USD 5.5 trillion.

Figure 2.1. **Investment needs in a 450* Scenario, 2014-35**

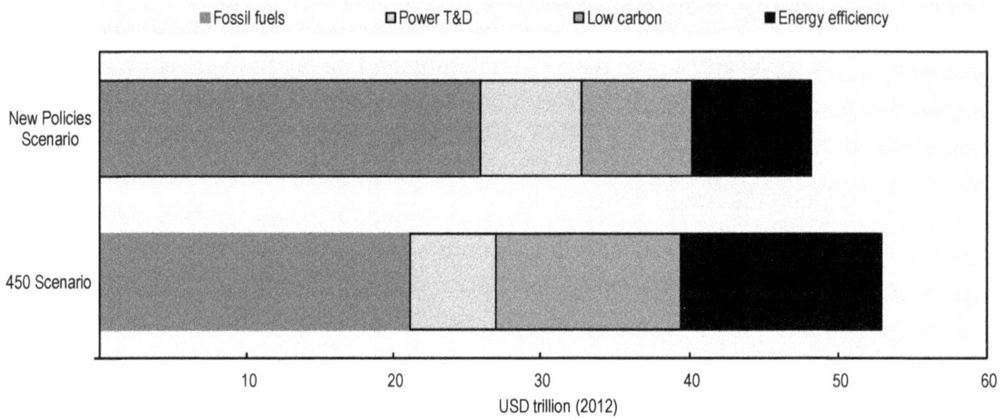

Notes: * The 450 scenario of the *Word Energy Outlook* (IEA, 2014) is compatible with a 2°C trajectory. Power T&D is transmission and distribution for the power sector.

Source: IEA (2014), *World Energy Investment Outlook*, Special Report, OECD/IEA, Paris, available at: www.iea.org/publications/freepublications/publication/WEIO2014.pdf.

There are different sources of traditional financing for infrastructure investments (IEA, 2014):

- Governments, companies and households can directly finance investments from their incomes: in companies, income that is not redistributed to shareholders is available for investment. Governments can directly invest in infrastructure through their budgets or through state-owned enterprises (SOEs).
- Banks and other financial institutions provide short- or longer-term loans to companies against an interest rate. Public financial institutions can provide loans and loan guarantee mechanisms in strategic areas to attract private sector investment.
- Capital markets provide a variety of long-term financing options though debt and equity. Bonds are the main debt instrument available on capital markets, issued by governments, public financial institutions or, to a lesser extent, companies. Equity usually attracts investors that are looking for higher risk.

The public sector has traditionally taken the lead in long-term investment in public goods, particularly in infrastructure projects. However, most OECD governments have had to tighten their budgets in the aftermath of the 2008 financial crisis; far fewer resources are available for public investment. Public investment per capita in 2012 fell in 15 out of 33 OECD countries, compared to 2007 (OECD, 2014a).

In light of the scale of the climate challenge and the growing fiscal constraints facing OECD countries, public financing alone will not be enough to meet these investments needs. The low-carbon transition will require the large-scale mobilisation of all sources of private sector investment and finance (Corfee-Morlot et al., 2012). Governments need to target their scarce resources at mobilising large-scale private sector investment and at activities unlikely to attract sufficient private funding.

Traditional sources of private capital, such as commercial banks, are facing increasing constraints on their ability to support long-term investment, with tightened financial regulations and the need to reduce debt. But banks are not the only source of financing – the low-carbon transition can mobilise new sources of capital and new financial instruments.

There is no shortage of capital in the economy. The total assets held by financial institutions – banks, institutional investors, central banks and public financial institutions – have been steadily increasing over the past ten years, amounting to around USD 305 trillion (FSB, 2013; see Figure 2.2). Not all of these funds are available for low-carbon infrastructure investments; central banks have specific mandates and purposes. However, the allocation of even a small fraction of these assets to low-carbon infrastructure would go a long way towards achieving the necessary low-carbon transition.

Figure 2.2. **Assets held by financial intermediaries in the 20 largest economies and euro area**

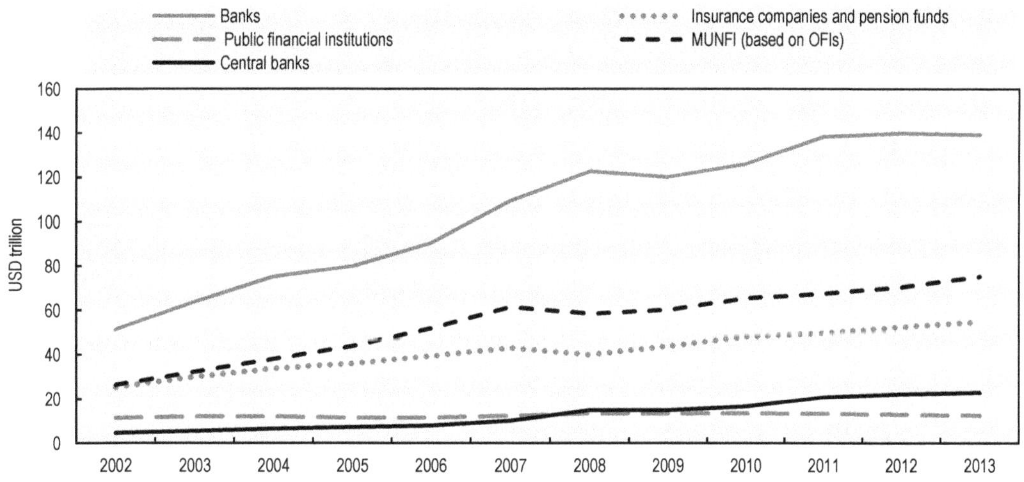

Notes: MUNFI: Monitoring Universe of Non-bank Financial Intermediation; OFIs: other financial institutions.

Source: FSB (2013), "Global shadow banking monitoring report 2013", Financial Stability Board, Basel, Switzerland, 14 November, available at: www.financialstabilityboard.org/wp-content/uploads/r_131114.pdf.

With USD 92 trillion of assets under management in OECD countries in 2013, institutional investors such as pension funds, insurers and sovereign wealth funds could play a significant role in driving long-term investments in a low-carbon economy. They are expected to increase in both scale and influence over the next decades due to greater wealth and the growing need to cater to ageing populations in both OECD and non-OECD countries. Those with long-term liabilities, such as pension funds, would be natural candidates to build broad portfolios of low-carbon investments, as they are looking for long-term, illiquid assets. Institutional investors have traditionally provided long-term capital with investment portfolios built around the two main asset classes (bonds and equities) and an investment horizon tied to the often long-term nature of their liabilities (OECD, 2015g).

Although institutional investors have increased their equity and debt investment in low-carbon projects in recent years, their investments remain minimal compared to the scale of their assets. Looking at large OECD pension funds only, direct investment in infrastructure projects of all types accounted for 1% of their asset allocation in 2013. Green infrastructure, including clean energy, is estimated to account for an even smaller share (ibid.). Too many barriers still stand in the way of scaling-up the participation of institutional investors (Box 2.2).

Overcoming barriers to shifting investment to low-carbon assets

Current levels of low-carbon investment fall short of a development pathway compatible with a low-carbon economy. Fossil fuels still account for more than 80% of final energy consumption, a share that has remained almost unchanged since 1990 (IEA, 2013). While investments in renewable energy have grown rapidly in the past ten years, from USD 60 billion in 2000 to a high point of nearly USD 300 billion in 2011, they have since fallen back, to USD 217 billion in 2013, and are dwarfed by investments in fossil fuels (BNEF, 2014). Of the USD 1 600 billion of global energy investments in 2013, 70% was in the extraction and transport of fossil fuels, oil refining and construction of fossil fuel power plants (IEA, 2014; see Figure 2.3). The same is true in the transport sector, where the majority of private investment in developing countries still supports road infrastructure (WRI, 2014b).

Figure 2.3. **Global annual energy supply investment, 2000-13**

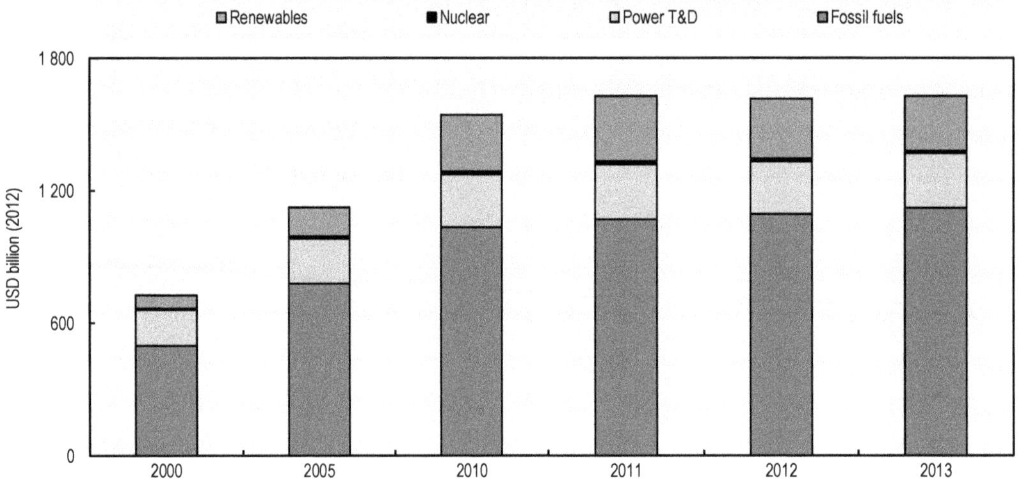

Notes: Power T&D is transmission and distribution for the power sector: this cannot be assigned either to fossil fuel or non-fossil fuel use.

Source: IEA (2014), *World Energy Investment Outlook,* Special Report, OECD/IEA, Paris, available at: www.iea.org/publications/freepublications/publication/WEIO2014.pdf.

Our economies encompass a range of market and policy failures that collectively favour investment in fossil fuel-intensive activities over investment in low-carbon infrastructure, often unintentionally. These are either linked to the enabling environment in specific sectors of the economy, or to the functioning and provisioning of financial markets (Table 2.1). Together they mean that the risk-return profile of low-carbon investments is often less attractive than their fossil fuel-based equivalents. Scaling-up financing to a low-carbon economy will not happen spontaneously. It requires strong policy and price signals to ensure that low-carbon and energy efficiency investments offer a sufficiently attractive risk-adjusted return compared to available alternatives (IEA, 2014).

The business environment is affected by policy misalignments in various domains. These are presented in the shaded part of Table 2.1 and are briefly discussed in the following section; they are the subject of other chapters in this report. A perspective on specific misalignments based on country context is also provided.

There are also policy obstacles embedded in financial systems and regulations that hinder the allocation of long-term finance to long-term low-carbon infrastructure investments (UNEP, 2014). Such barriers (listed in the non-shaded section of Table 2.1) include the way long-term investment is regulated, climate risk is valued, corporate outcomes are reported, public procurement operates and the way development finance is allocated and delivered. Removing these barriers will require some key architectural reforms to financial regulations, corporate governance and public spending, including development assistance policies. These areas are covered in the next three sections of this chapter.

Table 2.1. **Examples of policy misalignments that undermine low-carbon investment**

Business environment	Fiscal policies	– Insufficient carbon pricing and incentives for low-carbon technologies – Environmentally harmful subsidies and incentives (e.g. fossil fuels) – Tax policies that unintendedly favour carbon-intensive behaviour (e.g. company cars)
	Climate policies	– Lack of ambitious international and national reduction targets or binding objectives – Lack of climate policy stability: retroactive changes in climate legislation
	Investment policies	– Regulatory barriers to international investment in low-carbon projects (e.g. limits on foreign ownership, restricted access to land, local content requirements) – Lack of transparency, insufficient investor protection and intellectual property rights protection in low-carbon technologies, weak contract enforcement
	Competition policies	– Lack of open and competitive infrastructure markets (e.g. in the electricity sector) – Market designs and regulatory rigidities that favour carbon-intensive infrastructure investment in the energy sector – Lack of a level playing field in the power sector for existing fossil-fuel producing state-owned enterprises and independent producers of clean energy
	Trade policies	– Trade barriers for low-carbon goods and services
	Public governance	– Lack of long-term goals for low-carbon infrastructure planning and procurement – Contradictory signals between national and sub-national climate objectives – Lack of stakeholder consultation in policy design
Financial system	Financial market policies	– Potential unintended consequences of financial regulations on long-term financing – Financial incentives across the financial system favouring short-termism (remuneration practices, fiscal measures, performance appraisal) – Barriers to the deployment of innovative financial instruments for new types of investors (e.g. institutional investors)
	Business conduct	– Corporate reporting that does not reflect the climate risk (e.g. stranded assets) – Lack of a responsible investment code – Lack of clarity on fiduciary duty and stewardship with respect to environmental, social and governance issues
	Public finance and investment	– Ongoing support to carbon-intensive investments, nationally and internationally – Continued support of carbon-intensive investments in development finance – Lack of capacity

The enabling environment: Aligning investment policies with climate goals

Policy makers need to address policy misalignments in the overall investment framework that collectively favour investment in fossil fuel-intensive activities (identified in Table 2.1). These include conflicting incentives in competition, trade, tax and innovation policies, as well as inappropriate institutional settings. These are explored in this section, which builds on the key elements of the investment framework for green growth of the OECD Policy Framework for Investment[2] (OECD, 2015b).

There is no one-size-fits-all strategy when it comes to policy reform – the relative importance of each barrier depends on the country concerned. For instance, in developing economies the difficult business environment, lack of government capacity and low-maturity of financial markets are likely to be prominent obstacles to low-carbon investment. Broadening international financing avenues while supporting local financing mechanisms is essential to fill the financing gap (OECD, 2014b; 2013a). In OECD countries, the lack of stable and strong climate policies, unintended consequences of international financial regulations and environmentally harmful subsidies are likely to be more prominent (Table 2.2).

Elements of a policy framework for green investment

Strengthen government commitment to low carbon internationally and nationally

Uncertainty and expectations matter when considering returns on investments with long time horizons. Strong government commitments at both the international and national level are necessary to catalyse low-carbon green investment. With clear, long-term and ambitious signals and emission goals, nationally and internationally, investors and markets will have a clearer view on where to invest, which in turn could lead others to do so.

Many countries have already developed national climate laws and strategies, some of which are legally binding. The United Kingdom, for example, has introduced a long-term, legally binding absolute emissions reduction target in its Low-Carbon Transition Plan, translated into sequential carbon budgets.[3] Mexico's Climate Change Act (2012) also imposes domestic binding emission reduction targets. Ethiopia developed a green growth strategy that focuses on climate change and, by extension, food security (OECD, 2013a).

Conversely, confused and changing climate policies deter private investment and raise the cost of capital. Investors have repeatedly cited unpredictable changes to climate policies as a prohibitive barrier to low-carbon investment. Erratic changes in support schemes can also lower investor confidence and place further strain on the system (see Chapter 1).

Align investment incentives and disincentives to support a low-carbon agenda

Strong and stable carbon pricing policies or subsidies are needed to improve the returns on investment in green infrastructure projects and reflect the true long-term costs of economic development. Carbon pricing, regulatory interventions and robust support to not-yet-competitive low-carbon alternatives could provide consistent signals. Shifting investment incentives away from fossil fuels towards clean energy is also essential (see Chapter 1). In parallel to implementing a clear, predictable and long-term price on carbon, removing fossil fuel subsidies is fundamental for correcting distortions in the risk-return profiles of low-carbon investments (see Chapter 3 for a broader discussion on misalignments in the tax system).

Apply the principles of non-discrimination, transparency and property protection

Governments should also make sure that their investment frameworks for low-carbon investment are consistent with the essential investment principles of non-discriminatory treatment of cross-border investments, intellectual property and protection of technologies, and transparency. For example, local-content requirements are likely to increase the cost and slow the speed of market penetration by clean energy technologies, yet they have been planned or implemented in at least 21 OECD countries and emerging economies since the financial crisis (OECD, 2015c). Addressing outstanding barriers to international trade and investment in environmental goods, services and projects is also important (see Chapter 5).

Reinforce competition policies

Governments need to strengthen competition policy and address market and regulatory rigidities that favour incumbent fossil fuel and resource-intensive technologies and practices. Promoting competitive neutrality and keeping markets open will be key to maximising foreign and domestic investment, enhancing competitive pressure, promoting innovation and reducing costs. Preferential access to finance for outward investing and exporting state-owned enterprises can distort clean energy markets (OECD, 2015d). Ensuring a level playing field between incumbents and new entrants in low-carbon and climate-resilient infrastructure markets will be critical for the penetration of low-carbon technologies in the electricity sector (see opportunities for better alignment of innovation policies in Chapter 4, and electricity markets in Chapter 7).

Facilitate access to financing and attract co-financing for green projects

A wide range of suitable financial instruments (e.g. credit enhancement, leasing, guarantees, grants and bonds) offered by public financial institutions and by other companies, by investors via capital markets, and supported by an efficient financial infrastructure, are required to facilitate the transition to a resource-efficient economy (Cochran et al., 2014; Corfee-Morlot et al., 2012; OECD, 2015g). The role of public finance and public financial institutions is also critical, particularly in developing country contexts. Through consistency in their official development assistance (ODA) policies and broader development agendas, donor governments can help developing countries to improve the conditions for lower carbon development (see the last section of this chapter).

Enhance co-ordination and improve public governance at all levels of government

Early engagement by key stakeholders in goal setting and planning, at every level of government, will help ensure the relevance and consistency of policy objectives and expectations in markets. Enhanced co-ordination and improved public governance, especially among environment and natural resource management, energy and investment authorities, is also a critical factor for low-carbon investment. This will be particularly important at the metropolitan level to ensure a good integration of transport and spatial planning policies for low-carbon cities, or in the context of solving trade-offs in rural land use (see Chapter 8 on urban mobility and Chapter 9 on rural land use).

Establish policies to encourage environmentally responsible business conduct

Companies face increasing pressure to address climate change. Policy support to climate-friendly practices can leverage companies' contributions to addressing climate change over and above the influence of regulatory approaches such as carbon pricing. For instance, carbon corporate disclosure is an important element of responsible business, and should be increasingly encouraged by governments and investors (see section below on carbon disclosure).

Address other cross-cutting issues

Governments must also co-operate at the multilateral level to address barriers to international trade and investment in low-carbon infrastructure (see Chapter 5). More generally, there are a number of barriers preventing institutional investors from moving more into the low-carbon sector. Governments have a role to play in addressing these specific barriers (Box 2.2). In this regard, innovative financial instruments – such as green or climate bonds – could provide the missing link between long-term investments and a financial system increasingly focused on liquid, short-term securities and stocks (ibid.).

Country context matters

Each country has a distinctive constellation of misalignments in its investment framework. As in any investment project, the quality and resilience of political institutions and the legal system, lack of capacity locally and maturity of the financial system could be important obstacles to scaling-up private sector investment, and will influence the type of finance available as well as the required policy support. In addition, developing countries will face challenges, policy choices and trade-offs that are different from those of developed countries (OECD, 2013a; see also Table 2.2). Their different characteristics and priorities call for a specific sequencing and mix of policy instruments than from developed countries to achieve low-carbon and greener growth. For instance, the existence of a large informal economy that accounts for three-quarters of non-agricultural jobs in sub-Saharan Africa and over two-thirds in South and South-East Asia does not facilitate the implementation of economic, fiscal and regulatory policy instruments needed for a low-carbon transition (OECD, 2013a).

In this context, promising economic tools to promote low-carbon growth in developing countries include energy subsidy reforms with clear compensation mechanisms in place (see Chapter 1), environmental tax reforms in the context of a broader tax reform that could sustain the tax base (see Chapter 3) and payment for ecosystem services with accompanying capacity development measures (see Chapter 9). Implementing standards and certifications for sustainable production, mainstreaming sustainable public procurement and developing well-defined and transparent land tenure rights systems could further help developing countries integrate climate and a broader development agenda into economic decision making (OECD, 2013a).

The international community can help developing countries make the shift, especially by financing the short-term trade-offs that may be involved, such as the additional cost of going green. Given the scale of the challenge, however, much of the finance will need to come from domestic sources and international private investment, making it crucial to get the domestic enabling conditions right, including high-quality policy for low-carbon development. In its report *Putting Green Growth at the Heart of Development* (OECD, 2013a), the OECD proposes an agenda for international co-operation to support green growth that builds on three pillars: 1) strengthening international and domestic green finance and investment through better targeting all types of development finance (see below for a discussion on the need to mainstream climate in ODA and all development co-operation activities); 2) promoting green technology and co-operation, building capacity for domestic green innovation and adoption, and developing intellectual property rights regimes (see Chapter 4); 3) facilitating trade in green goods and services through the removal of tariff and non-tariff barriers (see Chapter 5).

Box 2.2. How to unlock institutional investors' investment in low-carbon, climate-resilient infrastructure

In addition to creating an adequate policy framework and other essential pre-conditions, there are a number of specific actions governments can take to overcome barriers and facilitate institutional investors' investment in green infrastructure:

- Build confidence. By developing a national infrastructure strategy, including a roadmap and project pipeline, investors would feel more confident in governments' commitment to green infrastructure.
- Develop liquid markets for green infrastructure financing instruments. These could be for debt in the form of green bonds, and for equity in the form of listed "yield co"-type funds.[1] They could be tailored to investor risk profiles across the project life cycle and developed in co-operation with investors.
- Create risk mitigation financing tools. Facilitating the development and application of risk mitigation financing tools could result in more appropriate allocation of risks and their associated returns. They could include credit enhancements and revenue guarantees, first-loss provisions, cornerstone stakes and tools targeting challenges at all stages of the project life cycle.
- Reduce transaction costs. The costs associated with managing green infrastructure investment could be eased by supporting efforts to standardise contracts and project evaluation structures, and by creating aggregation and "warehousing" facilities.
- Promote market transparency and standardisation. This could include improving data on performance, risks and costs of green investments across available channels while promoting public-private dialogue.
- Clarify the risk profile of green investments. Strengthening requirements for institutional investors to provide information on green investments, following internationally agreed definitions, would enhance monitoring and understanding of these investments.
- Be innovative in mobilising private investment. For example, a case can be made for establishing a special-purpose public "green investment bank" or refocusing activities of existing public financial institutions to mobilise private investment for green infrastructure.

Note: 1. A yield co is a publicly traded company that is formed to own operating assets that produce a predictable cash flow. Separating volatile activities (e.g. R&D, construction) from stable and less volatile cash flows of operating assets can lower the cost of capital. Yield cos are commonly used in the energy industry, particularly in renewable energy, to protect investors against regulatory changes.

Sources: OECD (2014a), "Annual Survey of Large Pension Funds and Public Pension Reserve Funds", OECD, Paris; OECD (2015g), *Mapping Channels to Mobilise Institutional Investment in Sustainable Energy*, Green Finance and Investment, OECD Publishing, Paris, http://dx.doi.org/10.1787/9789264224582-en.

Table 2.2. **How challenges and opportunities for low-carbon, resilient infrastructure investment vary by national circumstances or institutional context**

Category of country	Challenges	Opportunities
Developed countries	– Outdated or poorly maintained infrastructure requiring renovation – Lock-in: carbon-intensive infrastructure and urban development patterns – Climate policies in place but sometimes fragmented and partial – Some adaptation planning but limited policies and actions to adapt – Financial crisis put strains on banks and governments	– Renovation of aging infrastructure offers opportunities for upgrading to more climate-friendly infrastructure – Strong institutional capacity, certainty and reliability of the investment environment – High adaptive capacity, pockets of high vulnerability (e.g. urban slums) – High capacity to govern – New types of investors, such as institutional investors
Emerging economies	– Rapid economic development and severe environmental challenges, such as air pollution; inequalities – Medium adaptive capacity, relatively high and increasing vulnerability (e.g. slum populations in high-risk areas in cities) – Medium institutional capacity for policy reform, legal enforcement capacity – Rapid deployment of infrastructure, sometimes carbon-intensive (e.g. building fossil fuel-fired power plants to supply rapidly growing power demand) – Large state-owned enterprises with preferential access to finance may distort competition	– Relatively sound investment conditions and relatively well-developed capital markets – Strong economic growth and demographic pressure, rapid urbanisation, large investments in infrastructure occurring today – Strengthening institutional capacity and policies to address climate change – Rapid growth in new infrastructure provides opportunity to "leapfrog" technologically and integrate climate concerns at design phase at relatively low cost – Large policy-driven institutions, such as state-owned enterprises and public development banks (e.g. BNDES in Brazil)
Low-income and least developed countries	– Large informal economy – High level of poverty and inequality requiring targeted policies to avoid negative effects on the poorest – Weak capacity and resources for innovation and investment – Lack of basic infrastructure (e.g. transport, energy and water) and urgent need for economic development – High-dependence on natural resources (both renewable and non-renewable) – High poverty and high vulnerability to climate change and climate-related disasters (including slum populations in high-risk areas in cities) – Low adaptive capacity, some adaptation planning yet limited implementation or mainstreaming into development planning – Insufficient financial and technical capacity in government	– Growing international donor support for adaptation planning, mitigation implementation and technology transfer – Opportunities to integrate climate change consideration into development planning and infrastructure planning, which is largely led by the public sector – Low-carbon development can contribute to more sustainable management of the natural resources on which developing economies depend and reduce the pollution that can undermine long-term prosperity – Provision of basic infrastructure provides opportunity for leapfrogging; also where growth is limited and rural decentralised infrastructure solutions may deliver low-cost services (e.g. off-grid electricity) along with other local development benefits.

Sources: Adapted from OECD (2011), *Towards Green Growth*, OECD Green Growth Studies, OECD Publishing, Paris, http://dx.doi.org/10.1787/9789264111318-en; Corfee-Morlot, J. et al. (2012), "Towards a green investment policy framework: The case of low-carbon, climate-resilient infrastructure", *OECD Environment Working Papers*, No. 48, OECD Publishing, Paris, http://dx.doi.org/10.1787/5k8zth7s6s6d-en; OECD (2013a), *Putting Green Growth at the Heart of Development,* OECD Green Growth Studies, OECD Publishing, Paris, http://dx.doi.org/10.1787/9789264181144-en.

Channelling all sources of finance to low-carbon infrastructure

Governments need to better understand and monitor the potential unintended impact of rules governing the financial sector on the availability of public and private long-term finance for low-carbon projects. These include the way long-term investment is regulated, climate risk is valued, corporate outcomes are reported, public procurement operates, and development finance is allocated and delivered. Where negative unintended consequences are identified for climate, governments should evaluate the need to remedy them while making sure other fundamental development objectives, such as financial stability or inclusive growth, are met.

Ensure that the "rules of the game" for investor behaviour are consistent with a low-carbon pathway

In the aftermath of the 2008 financial crisis, governments have tightened the regulatory framework for banks at both the national and international level to discourage excessive risk-taking behaviour and increase the overall stability of the financial system. The Basel III accords, aimed at improving and harmonising the supervision and regulation of banks, strengthen the stringency of capital adequacy requirements.

Some argue that Basel III is having unintended consequences on the ability of private financial actors to invest in and finance low-carbon, climate-resilient infrastructure. The concern in the energy sector in particular is that the capital and liquidity requirements of Basel III may limit the amount of capital available for long-term financing from banks in the future (IEA, 2014; Narbel, 2013; Spencer and Stevenson, 2013; UNEP, 2014, 2015).

There is to date a lack of data and empirical evidence to support these claims. The Financial Stability Board (FSB) has been mandated by the G20 to monitor financial regulatory factors affecting the supply of long-term investment finance. It "finds little tangible evidence or data to suggest that global financial regulatory reforms have had adverse consequences on long-term investment" (FSB, 2012). Many FSB members stress that it is too early to fully assess the effects of reforms still in the early stages of implementation or being developed, as Basel III only comes into full force in 2019. It can also be very difficult to disentangle effects of regulatory changes from broader economic and policy factors that affect supply and demand of long-term investment finance.

Solvency II is also cited as barrier to investors' involvement in the sector, affecting European insurance companies in particular. Like Basel III, Solvency II introduces more stringent quantitative (solvency ratios) and qualitative (risk management and supervision) requirements for European insurance companies. It increases the quantity and quality of regulatory capital that insurers have to hold to cover their insurance and investment risk. The regulation is said to have the effect of inducing insurers to reallocate investments away from equity – as equity investments are subject to a higher charge than debt – towards more highly rated securities. The impact on direct infrastructure investment has also been cited as potentially negative, as Solvency II no longer limits the amount that insurers may invest in non-listed assets (such as infrastructure, private equity, venture capital; Spencer and Stevenson, 2013). Solvency II's calibrations are currently under review.

Financial stability is a prerequisite to any kind of investment, including low-carbon ones. However, governments need to continue to monitor financial regulatory reform for unintended consequences on the supply of long-term investment financing, and should ascertain the extent to which these regulations are consistent with the objective of transitioning to a low-carbon economy. Such regulations include financial regulation but also accounting practices and other specific regulations to institutional investors on the availability of long-term investment finance.

Properly account for climate risks and liabilities in financial markets

While in general financial markets are perceived to be fairly efficient, mispricing of risk can still occur. For example, there is growing concern that the market risk related to climate liabilities embedded in corporations' assets or investors' portfolios is neither properly assessed nor managed. Stronger climate policies will have an impact on future investment decisions, but also on the profitability of existing assets. Some existing fossil fuel assets might not be able to fully recover their investment due to pricing and profitability changes resulting from climate change regulation. Properly accounting for this risk in financial evaluations could help better allocate capital to low-carbon, climate-resilient assets, or even encourage divestment from more carbon-intensive assets.

Stronger climate policies will have an impact on the profitability of existing assets. There are several degrees of climate-related risks that could create "stranded assets", i.e. premature devaluation of assets or liabilities in investors' and corporations' portfolios (Carbon Tracker Initiative, 2013):

- Regulatory stranding due to a change in climate legislation, e.g. an evolution in minimum standards for greenhouse gas (GHG) emissions reductions from thermal power plants.
- Economic stranding due to a change in relative costs and prices, e.g. through the implementation of a significant carbon price across the economy, or because of a decline in overall energy prices. The effects of the recent drop in oil prices on investors' portfolios are a striking example of this phenomenon.
- Physical stranding due to extreme weather events, or gradual changes that hinder the operation of assets (see Chapter 6).

For the energy sector alone, the *World Energy Investment Outlook* (IEA, 2014) estimates that by 2050, USD 300 billion in assets could be stranded under a scenario compatible with the 2°C objective:

- Power sector: 165 gigawatts (GW) of new fossil fuel capacity would have to be retired before repaying investment costs, with an unrecovered sunk cost of USD 120 billion. Ninety GW of new power plants would go through early retirement, but after having recovered their investment costs.
- Oil and gas: 5% of proven oil and gas reserves would be stranded, representing USD 130 billion for oil and USD 50 billion for gas. These reserves are often referred to as "unburnable carbon".
- Coal: here the risk of stranding assets is relatively low, at USD 4 billion, as the low-carbon projection implies the closure of older, fully repaid mines.

To date, this risk is not properly disclosed, let alone priced, across investors' or governments' portfolios. As part of their commitment to climate change mitigation, governments, regulators and central banks should start addressing this market failure alongside the implementation of core climate policy instruments. Central banks and other financial regulators, if they have authority, may need to take action to examine the risks that climate change poses to the real economy. Stock exchange regulations could, for instance, demand more precise disclosures of carbon content in listings.

Failure or delay to create a clear and stable policy framework could increase the risk of stranding, as more "dirty" assets would be built only to be shut down or operated at a level much lower than anticipated by investors. This would increase the overall costs of implementing ambitious climate policy. In addition, governments could themselves be financially affected by stranded assets, as they own 50-70% of global oil, gas and coal resources and collect taxes and royalties on portions they do not own (CPI, 2014).

Public and private investors should increasingly assess climate risks. The *G20/OECD High-Level Principles of Long-Term Investment Financing by Institutional Investors* state that "the risks associated with long-term investments should also be carefully assessed, including market and illiquidity risks (and related portfolio constraints), climate and other environmental risks, and exposure to potential future climate regulation" (G20/OECD,

2014). The Bank of England is to conduct an enquiry into the risk of fossil fuel companies causing a major risk to financial stability if future climate change rules render their coal, oil and gas assets worthless.

Enhancing climate risk disclosure by corporations and investors

If companies were able to better measure and be more transparent about their liabilities related to climate change, it would allow investors to evaluate assets more accurately, increase social accountability and facilitate the design of incentives to encourage climate-friendly behaviour. Various elements of environmental, social and governance (ESG) disclosure could help to lower the vulnerability of companies to climate policy and to climate change risks (i.e. stranding or actual destruction of assets).

Step up standards for environmental, social and governance disclosure

ESG corporate disclosure is an important element of responsible business and is increasingly in demand by governments and other stakeholders. Corporate disclosure practices are increasingly supported (or required) by legislation or stock exchange regulations that make ESG disclosure mandatory. Experience shows that for companies above a certain size and complexity, ESG disclosure can be an important tool for identifying business risks and opportunities (Baron, 2014). ESG disclosure usually includes GHG reporting, which is a first necessary step in creating financial indices that include information on carbon emissions. Indices such as the *Low Carbon 100 Europe* or the *FTSE4 Good Index Series* permit dedicated tracker funds to offer investment opportunities for investors interested in financial products with an emissions metric.

Since the late 1990s, a number of mandatory or voluntary government schemes have emerged that require or encourage enterprises to measure and report their GHG emissions. The majority of G20 countries now have some kind of corporate reporting scheme in place that requires disclosure of climate change-related information (Box 2.3).

Box 2.3. Current status of corporate reporting legislation internationally

In the **European Union**, the EU Emissions Trading System (ETS) covers companies in energy-intensive sectors, including more than 11 000 power stations and manufacturing plants in the 28 EU member states and other European Economic Area countries. In total, around 45% of total EU emissions are covered by the EU ETS (EC, 2013). Installations are required to measure direct emissions each year, and provide emissions reports verified by an accredited verifier. In addition, the EU Directive on financial reporting was amended in 2014 to require large public interest entities with more than 500 employees to also report on non-financial information (EC, 2014). Reporting requirements include disclosure on policies, outcomes and risks, and relevant non-financial key performance indicators concerning environmental and social matters, human rights, anti-corruption and bribery issues, and diversity of directors. The Directive will apply to approximately 6 000 EU entities (up from 2 500 companies currently reporting). The amendment came into force in 2014; national governments have two years to incorporate it into national law. The first corporate reports under the scheme will relate to the financial year 2017.

Box 2.3. Current status of corporate reporting legislation internationally (*cont.*)

The **United Kingdom** moved from voluntary to mandatory reporting for GHG emissions in 2013. Under this scheme, all UK quoted companies must report on their GHG emissions as part of their annual directors' reports. The requirement affects all UK incorporated companies listed on the main London Stock Exchange, another European market or whose shares are dealing on the New York Stock Exchange or NASDAQ.

Under **France**'s 2010 Grenelle II legislation, companies with over 500 employees are required to publish a carbon inventory and an emissions reduction, and to include report information on a broad range of environmental social and governance related issues in their management plan. As with the United Kingdom, France requires this information to be third-party verified.

In the **United States**, the Environmental Protection Agency (EPA) issued the Mandatory Reporting of Greenhouse Gases Rule in September 2009. The rule governs mandatory reporting of GHG for suppliers of fossil fuels or industrial GHGs, manufacturers of vehicles and engines, and all facilities that emit more than 25 000 tonnes of GHG every year (US EPA, 2014). In 2014, the EPA's Greenhouse Gas Reporting Program released its 4th year of emissions data, including information from facilities in 41 source categories. In 2010, the Securities and Exchange Commission issued interpretive guidance on climate change disclosure, but so far it seems to have had limited effects: 59% of Standard & Poor's 500 companies do report on climate but, according to some, their disclosures have been disappointing (Ceres, 2014).

In **Australia**, the National Greenhouse and Energy Reporting Act of 2007 introduced a national framework for the reporting and dissemination of information on GHG emissions, GHG projects, energy use and production by companies. This reporting requirement has not been changed in spite of the July 2014 repeal of the 2011 Climate Tax.

Japan has a long history of regulation and incentives to reduce energy consumption and increase energy efficiency. GHG emission is part of the system and a range of governmental schemes are in place to support energy and climate change policies, including the 2006 Mandatory Greenhouse Gas Accounting and Reporting System, Japan's Voluntary Emissions Trading Scheme and the Experimental Emissions Trading Scheme.

Carbon reporting is also required in emerging economies

The **People's Republic of China**'s (hereafter "China") National Development and Reform Commission (NDRC) has developed GHG Accounting and Reporting Guidelines for ten industries, using the GHG Protocol (WRI, 2014a). In March 2014, the NDRC announced a new regulation requiring all firms emitting more than 13 000 tonnes (CO_2-eq) to begin reporting their annual GHG emissions. It is expected to be enforced from 2015 (Reuters, 2014).

In **India** the National Voluntary Guidelines on Social, Environmental and Economic Responsibilities of Business, adopted in 2011, encourage companies to disclose non-financial information, but no specific GHG emissions reporting scheme is currently in place. In 2013, the India GHG Program was established by Indian companies and think tanks in co-operation with the World Resources Institute to increase the capacity of Indian companies in measuring and disclosing GHG emissions (India GHG Program, 2014).

In its White Paper on a Climate Change Response, **South Africa** is planning to make carbon reporting mandatory by large emitters: those emitting over 0.1 Mt of GHGs annually or that consume electricity which results in more than 0.1 Mt of emissions from the electricity sector.

In parallel, a range of tools (principles, guidelines or certification-based standards) now exists to facilitate ESG reporting by companies (Baron, 2014):

- The Global Reporting Initiative guidelines: a leading ESG reporting framework providing a clear structure and set of metrics, including on GHG emissions, that companies can apply and adjust to their business areas (GRI, 2013).
- The Climate Disclosure Standards Board: designed to help companies disclose information about their climate change-related risks and opportunities, carbon footprints, carbon reduction strategies and implications for shareholder value in mainstream financial reports (CDSB, 2013).
- The Climate Disclosure Project: the largest organisation collecting voluntary corporate reports on climate data. Over 11 000 companies worldwide reported to it in 2013 (CDP, 2014).

Ensure climate disclosures provide a full view of a company's future in the face of climate change risks

Although progress has been made in corporate ESG disclosure, there is clearly room for more, including on climate. While homogenisation is growing via the widespread use of the Global Reporting Initiative guidelines, corporate reports are still difficult to compare, increasingly weighty (sometimes running to hundreds of pages) and not always satisfactory in terms of the quality of information provided (WBCSD, 2013). In a survey conducted by the Association of Chartered Certified Accountants to feed into the European discussion on mandatory corporate reporting of non-financial elements (ACCA and Eurosif, 2013), 92% of investors surveyed found the information provided by companies not sufficiently comparable.

ESG reporting is not internationally harmonised. It exists in some jurisdictions, but is voluntary or non-existent in others. With the exception of Denmark, the Netherlands and the United Kingdom, board responsibility is not mandated in most countries. Some carbon disclosure schemes cover the largest GHG emitters; others cover the largest firms (e.g. in terms of employees) but leave out a large majority of companies. Some focus on emissions, others also cover climate risks. Some schemes provide a reporting framework, others leave latitude in the choice of a reporting framework and indicators.

Another major critique, certainly in relation to climate-related information, is a lack of clarity about what is and is not material to a company's business, which limits the usefulness of a company's report for investors. For instance, Ernst and Young (2014) found that institutional investors are often unable to identify what issues presented in ESG disclosures could materially affect shareholder returns. Similarly, investors have difficulty connecting non-financial and financial performance, and comparing across companies.

Use integrated reporting as a framework for climate disclosure

A better alignment of corporate reporting schemes with the low-carbon transition would require greater harmonisation of ESG reporting requirements and greater coherence among scope, risk disclosure and reporting requirements on climate-related information while minimising bureaucratic burdens for small and medium-sized enterprises and avoiding distorting competition.

More fundamentally, it is argued that climate disclosures do not provide a full view of a company's future in the face of climate change risks, let alone broader social and environmental issues. For instance, the scope of climate-related information to be disclosed under government schemes is generally limited to GHG emissions, and only a few require reporting of emissions reduction targets or climate risk management plans.

The Carbon Tracker Initiative (2014) provides a list of disclosures that would increase, in their view, fossil fuel companies' transparency about climate risks. These include scenario analyses and more direct measurements, e.g. of carbon embedded in coal, oil and gas reserves and resources. Reports on the vulnerability of assets to climate change impacts can also be relevant. However, they also raise questions about comparability and methodologies used to measure and report vulnerability. Are companies relying on the same baseline climate change scenarios? How are extreme events treated? If a materiality filter is applied, how are various risks evaluated (is probability of occurrence factored in, or should a company be reporting on risks even in the absence of a probability function)?

Looking further ahead, the end goal seems relatively clear: concise corporate disclosure that links financial and ESG performance, both to foster action in companies that have so far ignored climate change risks and may be at risk as a result, and to encourage investments in companies with highly integrated performance.

This is the goal pursued by companies supporting "integrated reporting", the disclosure by a company of information on its near-, medium- and long-term capacity to generate value, including material risks and opportunities related to all its capitals: financial, manufactured, intellectual, human, social and relationship, and natural (IIRC, 2013). While a common statistical basis is still needed, good practice is emerging: integrated reports are now mandatory for listed companies in South Africa, and in France publicly listed and other companies must issue a report combining financial and ESG information.

Mainstreaming climate change goals in public spending

Domestic and international public financial institutions and development agencies can be instrumental in providing risk mitigation instruments and refinancing guarantees to bridge the long-term financing gap, provided they have a clear mandate and mainstream climate change into their economic decisions.

Mobilise public financial institutions

Public financial institutions, including national banks, development agencies and export credit agencies, are all engaged in decisions that could either lock in more carbon or facilitate the transition away from fossil fuel use. As policy-driven institutions, they were originally created to address market failures or externalities which limit private sector investment, and to deliver financial services that meet a public policy objective and are not provided by the market. In some cases, these institutions are mandated to provide long-term financing independent of market cycles and in line with policy priorities; they are also able to leverage capital at advantageous, below-market rates for targeted investments, as they benefit from government support (Cochran et al., 2014).

These characteristics and objectives of public financial institutions fit well with the need to step-up private investment in low-carbon projects. Public financial institutions can be instrumental on three levels:

1. facilitating access to long-term financing
2. reducing project and financial risks

3. filling the capacity gap (i.e. providing expertise to support low-carbon investments and market development).

Some PFIs already have an explicit mandate and authority to invest in green infrastructure (e.g. KfW) – often with established guidelines on which technologies or markets to address. Others have a much less precise mandate, with "sustainability" being part of their overarching objectives (Cochran et al., 2014). Export credit agencies have also started exploring to what extent they could support the climate agenda (Box 2.4).

Box 2.4. The role of export credit agencies in the low-carbon transition

The main objective of export credit agencies' (ECAs) is the promotion of trade. ECAs are demand-driven public institutions that provide funds and guarantees to domestic companies to facilitate exports, including for fossil fuel investments. Export credit terms are regulated by the OECD Arrangement on Officially Supported Export Credits.

OECD countries have taken active steps to introduce and maintain environmental accountability in official export credits and to address climate change. Following the introduction of a Sector Understanding on Export Credits, Renewable Energies and Water Projects for a trial period in December 2005, participants agreed in 2012 to a Sector Understanding on Renewable Energy, Climate Change Mitigation and Water Projects, to complement the arrangement and provide adequate financial terms and conditions to projects in sectors significantly contributing to climate change mitigation, including renewable energy and energy efficiency projects.

In 2014, OECD member countries agreed a Ministerial Statement on Climate Change, which affirmed members' commitment to "continuing discussions on how export credits can contribute to our common goal to address climate change". Discussions are ongoing in the OECD Working Party on Export Credits on how export credits might help address climate issues.

Source: OECD (2012b), "Arrangement on officially supported export credits", TAD/PG(2012)9, OECD, Paris, available at: www.oecd.org/officialdocuments/publicdisplaydocumentpdf/?doclanguage=en&cote=tad/pg(2012)9.

In the last ten years, countries have complemented their mainstreaming efforts by establishing special purpose banks to finance the low-carbon transition. Known as green investment banks,[4] these are "domestically focused public institutions that use limited public capital to leverage or 'crowd-in' private capital, including from institutional investors, for low-carbon and climate resilient infrastructure investment" (Eklin et al., 2014). They are hence scaling-up private investment in green infrastructure, and creating a track record for investment in clean energy. While these efforts should be encouraged, all public financial institutions should mainstream climate change aspects in their investment decisions.

Strengthening the climate mandate of public financial institutions would facilitate the low-carbon transition. In particular, public banks could play an important role in supporting the refinancing required for long-term low-carbon projects, as traditional commercial banks are less willing to lend long term. Public banks could develop low-carbon refinancing guarantee facilities to overcome this bias (Spencer and Stevenson, 2013).

Make use of green public procurement and expenditure

Public procurement is essential to the commissioning and delivery of infrastructure projects. General government procurement accounts for 13% of gross domestic product (GDP) and nearly one-third of government expenditures in OECD countries (OECD, 2013b). As low-carbon projects can be more expensive than alternatives, especially if carbon emissions are not explicitly valued, there is no incentive for the private sector to include them in public bidding processes unless it is required (Corfee-Morlot et al., 2012). This is where green procurement policies come in. Sustainable public procurement could shape consumption and production to support low-carbon objectives, generate new markets and provide examples of good practice for business and consumers (OECD, 2013a).

OECD countries increasingly recognise how procurement can be used for different objectives, not only value for money and integrity, but also wider objectives such as sustainable development and greening public infrastructure. Green public procurement is defined in the OECD *Recommendation of the Council on Improving the Environmental Performance of Public Procurement* (OECD, 2002) as "the procurement of products and services which are less environmentally damaging when taking into account their whole life cycle". The OECD further adopted the *Recommendation of the Council on Public Procurement* (OECD, 2015e) to encourage the use of procurement as a smart governance tool, including recommendations on how to integrate secondary policy objectives, such as sustainable green growth. Today, 72% of OECD countries already have policies encouraging green procurement at the central government level. Some developing countries are starting to adopt green procurement practices: Colombia has attached a green criterion to procurement policy including life-cycle analysis of products (OECD, 2013a).

To contribute to the OECD Green Growth Strategy and provide concrete examples for countries wishing to invest in green public procurement, the OECD's Public Governance Committee has developed a compendium of good practices (OECD, 2015f; see also Box 2.5). The compendium presents case studies across six dimensions which correspond to green public procurement challenges reported by countries to the OECD.

Challenges include procurement officials' perceptions that green products and services are more expensive than conventional ones, lack of technical knowledge on how to integrate environmental standards in the procurement process, challenges in accurate life-cycle costing and the absence of reliable monitoring mechanisms to evaluate if green public procurement achieves its goals.[5]

Governments should encourage the use of procurement to ensure effective public service delivery while pursuing the goal of a low-carbon economy at all levels of government. To do so, the low-carbon objective of procurement should be clearly articulated and prioritised. Consideration should be given to which means are best able to achieve this objective: traditional procurement, public-private partnerships (PPPs), concessions, etc. In addition, systematic life-cycle analysis should be undertaken to properly account for the costs and benefits of low-carbon procurement. While 79% of OECD countries identify the cost of green projects as a key barrier to further mainstreaming green procurement, only 16% of countries implement a life-cycle cost evaluation systematically (Box 2.5).

Box 2.5. Compendium of green procurement good practices

The compendium of green procurement good practices aims at helping countries implement green public procurement (GPP) across six areas:

1. **GPP legal and policy framework.** Certain OECD countries such as Germany, Japan and the United States have put in place a clear legal framework which has allowed them to direct purchasing activities to achieve set green goals. Non-OECD member countries such as Colombia and China are moving fast to put GPP policies in place.
2. **Planning GPP, assessing life-cycle costs and understanding market solutions and capacity.** Only 16% of OECD countries evaluate life-cycle costs systematically when purchasing. Such analysis should be systematically mainstreamed into green procurement practices. It is also important to understand what green solutions are available and industry appetite to provide green products, otherwise tender procedures may fail.
3. **Environmental standards in the design, selection and award of projects and contract performance.** In 2010, 24 OECD countries included environmental considerations in technical specifications for products and 18 in the award criteria for contracts, but only 13 in contract performance, for example with a bonus if a contractor achieves or exceeds certain standards in performance.
4. **Professionalisation, multidisciplinary procurement teams and GPP training.** Multidisciplinary means teams that include procurement officials, lawyers and professionals with technical GPP capacity, such as engineers. Procurement knowledge and skills are essential. In the case of GPP, which adds the environmental angle to all other procurement complexities, bureaucratic and otherwise, this is even more relevant.
5. **Raising the awareness of buyers, the market and citizens of GPP solutions and benefits.** There is little data on awareness campaigns and communication strategies undertaken in OECD countries to raise the visibility of GPP.
6. **Mechanisms to monitor the impact of green procurement.**

Source: OECD (2015f), "Smart procurement: Going green – Best practices for green procurement", GOV/PGC/ETH(2014)1/REV1, OECD, Paris, available at: www.oecd.org/officialdocuments/publicdisplaydocumentpdf/?cote=GOV/PGC/ETH%282014%291/REV1&docLanguage=En.

Improve the use of discount rates in economic valuation

The use of high discount rates in the financial sector can exclude resilience and values of natural and social capital from financial investment decisions and arbitrages. Time horizons of investments, even those labelled as long-term, do not match those of low-carbon and climate resilience investments. This is true of public investments also. Governments should review their use of discount rates to ensure they are not resulting in bias towards carbon-intensive infrastructure. The use of unique discount rates in public project evaluation can create a bias, as the positive externalities of low-carbon projects are often long-term and could be implicitly excluded from the valuation. More work is needed to investigate best practices for project valuation mechanisms at the government level, including the incorporation of a proxy for the monetised value of externalities (see Chapter 3). More work is also needed to understand how to better align time horizons from public and private sector perspectives.

Mainstream climate change across all development co-operation policies

Development and environment, including climate change, are inextricably linked: without further policy action, local and global climate risks threaten to reverse development gains made to date by exacerbating water, food and other resource scarcity, as well as increasing the risk of extreme weather disasters (OECD, 2014b).

Development finance for climate change is on the increase, largely driven by international commitments and financial mechanisms under the Rio conventions (Figure 2.4).[6] Total bilateral and multilateral climate-related development finance commitments to developing countries reached USD 37 billion in 2013.[7] This represents up to 17% of bilateral official development assistance (ODA) and 19% of multilateral development finance flows in 2013 (OECD-DAC Statistics, 2014). The majority of finance targets climate change mitigation objectives, with USD 23 billion (74% of the USD 37 billion) targeting mitigation only or mitigation together with adaptation objectives, while USD 9.6 billion (39%) targeted adaptation (either alone or together with mitigation activities). Key infrastructure sectors – energy, transport and water – received over two-thirds of climate-related development finance in 2013.

Bilateral climate-related ODA commitments from members of the OECD Development Assistance Committee (DAC) have been rising, reaching USD 21.9 billion in 2013 (Figure 2.4). Of this, USD 12.4 billion targeted climate change as a principal objective, meaning that these development co-operation activities would not have taken place in the absence of a climate change objective (OECD-DAC Statistics, 2014).

Figure 2.4. **Trends in bilateral official development assistance to climate, 2002-13**

3-year annual averages

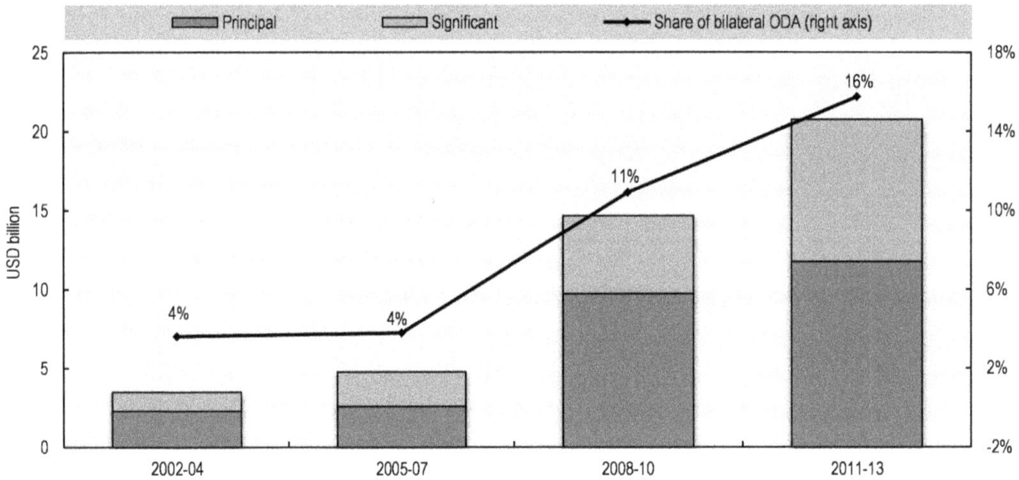

Notes: The figure presents a trend based on averages over three years to smooth fluctuations from large multi-year projects programmed and committed in a given year, such as observed in 2010. Development co-operation to adaptation activities was only measured from 2010 onwards, so this aspect is underestimated before that point. Activities marked as having a "principal" climate objective (mitigation or adaptation) would not have been funded but for that objective; activities marked "significant" have other primary objectives but have been formulated or adjusted to help meet climate concerns.

Source: OECD DAC Statistics (2014), *Climate-related development finance in 2013: Improving the statistical picture,* OECD, November 2014. www.oecd.org/dac/environment-development/Climate-related%20development%20finance%20FINAL.pdf.

While climate-focused development finance has been steadily increasing, much more can be done to further integrate and mainstream climate change across all bilateral and multilateral development co-operation providers and sectors. To date, less than 20% of development finance targets climate change objectives. There is no guarantee that the 80% of ODA that is not targeted to climate is not increasing GHG emissions or climate vulnerability. Assessments suggest that a sizeable share of development assistance activities might be affected by climate risk, with estimates ranging from 10-40% per country depending on the development co-operation portfolio in each country context, when measured as a share of total ODA .

This is particularly important as the share of ODA targeting infrastructure is increasing, with a sizeable portion disbursed to support private sector investment through loans and equity by bilateral and multilateral institutions (Miyamoto and Biousse, 2014). In the energy sector, bilateral development finance (ODA and other official flows, or OOF) nearly doubled, from USD 5.4 billion a year over 2002-04 to USD 10.3 billion a year over 2011-13. Within this, development finance targeting renewable and nuclear energy has also grown steadily, and at a faster rate – peaking at 40% in 2008-10 before falling slightly to 35% in 2011-13 (Figure 2.5). While the share of development finance to non-renewable energy has fallen steadily, finance to non-renewable energy did increase between 2008-10 and 2011-13 in absolute value.

Figure 2.5. **Bilateral development finance* to the energy sector, 2002-13**

3-year averages

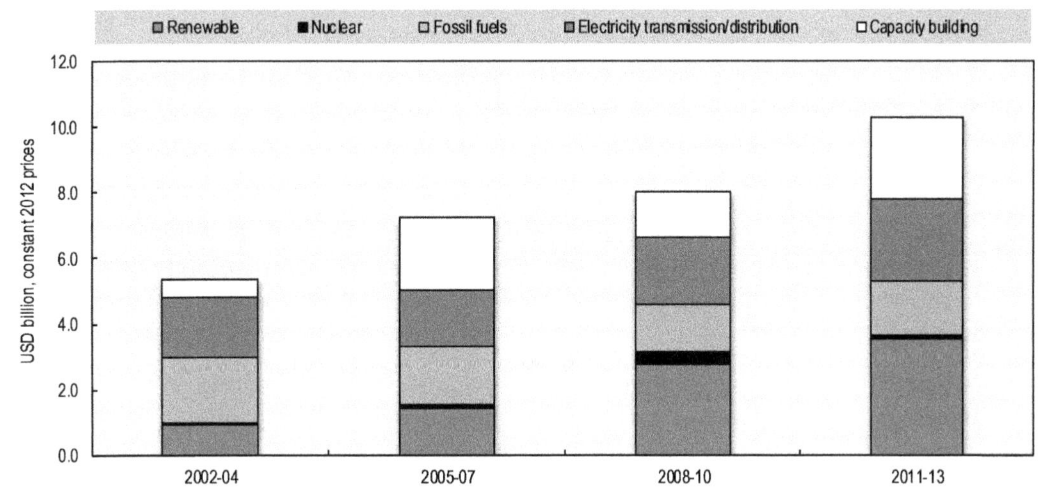

Notes: *Development finance includes official development assistance and "other official flows" (transactions by the official sector which are not eligible as official development assistance, either because they are not primarily aimed at development or because they have a grant element of less than 25%).

"Renewable" encompasses biomass, geothermal energy, hydro-electric power plants, ocean power, solar energy, wind power and power generation from renewable sources.

Source: OECD-DAC Creditor Reporting System, December 2014.

Traditional development co-operation providers need – and are increasingly aiming – to be smarter in their use of development finance to ensure that the private sector and private finance are mobilised towards developing a low-carbon economy. A recent analysis of ODA support to private investment in infrastructure shows that 58% of support to private sector energy investment goes to renewable energy sectors (Figure 2.6; Miyamoto and Biousse, 2014). Beyond renewable energy, no reference to low-carbon and climate-resilient infrastructure in the transport sector is found in donors' strategies for supporting the private sector, except for the European Bank for Reconstruction and Development (EBRD), which has a specific focus on environmentally sustainable transport.

Figure 2.6. **Sectoral distribution of official development assistance support to private investment in infrastructure, 2011**

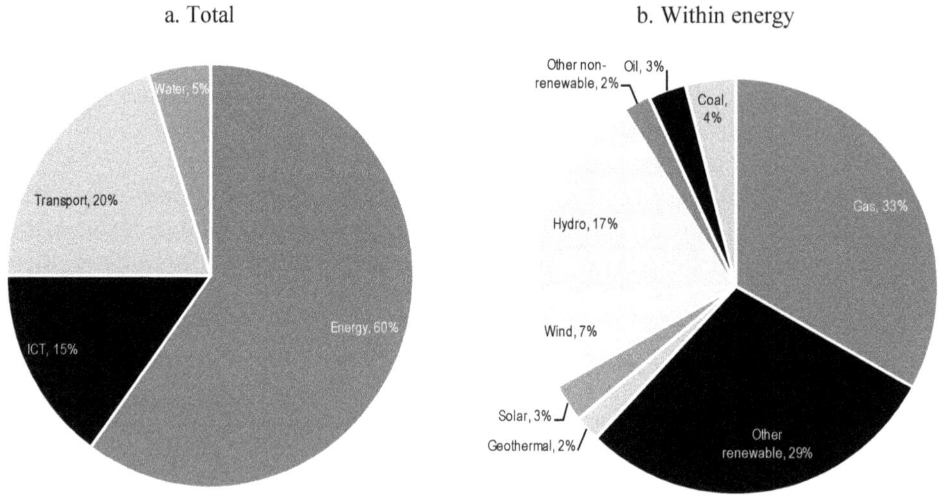

Source: Miyamoto, K. and K. Biousse (2014), "Official support for private sector participation in developing country infrastructure", *OECD Development Co-operation Working Papers*, No. 19, OECD Publishing, Paris, http://dx.doi.org/10.1787/5jz14cd40nf0-en.

Middle-income countries receive 80% of climate mitigation-related development finance. These are where most mitigation leverage exists, and well-developed financial markets create good conditions for mobilising the private sector. However, it does raise the question of whether in this case ODA may be substituting for private investment (Miyamoto and Biousse, 2014).

More generally, greater integration and mainstreaming of climate change considerations across development co-operation activities requires a better understanding of the new development finance landscape, with new actors, geographies and forms of innovative financial support. The traditional image of donor countries and aid recipients is now out of date.

Development finance is now much more dynamic and increasingly driven by new providers participating in South-South co-operation and engaging with provider countries in triangular co-operation. Private sources of finance in the form of equity, bonds, loans and risk mitigation instruments, as well as philanthropic sources, could become transformative agents (OECD, 2014e). ODA only represents 6% of external finance in many upper middle-income countries; here ODA could move away from traditional concessional finance to the use of credit enhancement mechanisms such as guarantees, hybrid debt/equity, insurance schemes and securitisation (OECD, 2014e). Many large emerging economies have well-developed financial systems, with stock exchanges and debt markets that help them mobilise domestic and international private investments (Figure 2.7). For least developed countries, however, ODA still represents 70% of their total external finance as grants, and is still very important.

Figure 2.7. **Developing countries' net resource receipts from DAC countries and multilateral organisations in 2000-11**

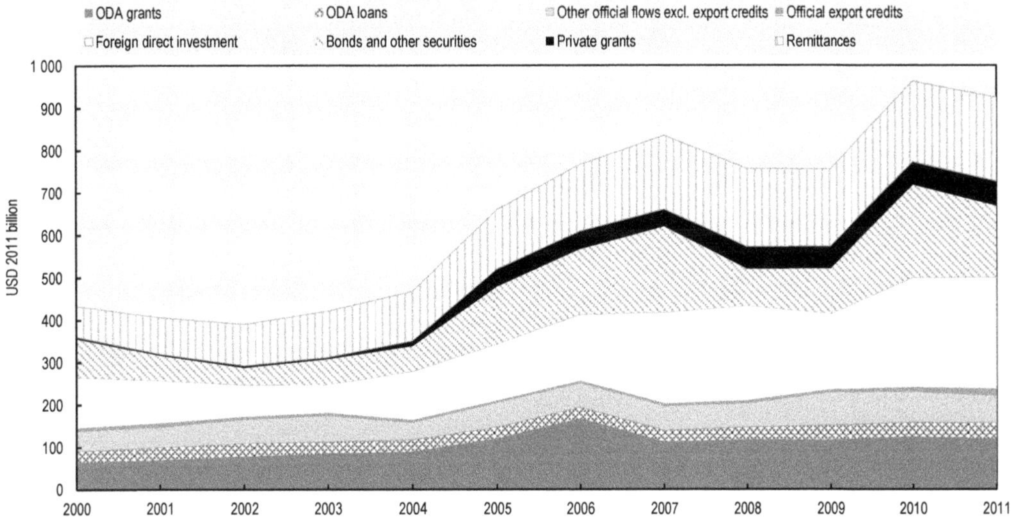

Notes: Total external financial resources include bilateral official development assistance (ODA), other official flows (OOF), private grants, private flows at market terms and remittances from Development Assistance Committee (DAC) member countries, and concessional and non-concessional outflows from multilateral agencies. Since 2005, private grants have been based on estimates from the Hudson Institute's Center for Global Prosperity, which uses a more generous definition than DAC statistics, including, for example, the imputed value of volunteer time.

Source: OECD (2014b), *Development Co-operation Report 2014: Mobilising Resources for Sustainable Development*, OECD Publishing, Paris, http://dx.doi.org/10.1787/dcr-2014-en.

The lack of data is an obstacle to enhancing the effectiveness of climate finance. At the international level, while significant progress has been made in monitoring and reporting public flows, there is no integrated measurement, reporting and verification system for all climate-specific flows, as they come from various sources (North, South), through bilateral or multilateral channels and various instruments (offset finance, grants, concessional loans, capital, guarantees, risk mitigation tools). One very specific data challenge is the tracking of climate finance linked to international negotiations on climate change under the United Nations Framework Convention on Climate Change (UNFCCC). At the UNFCCC Conference of the Parties in Cancun in 2010 (COP16), member states:

- recognised that developed country parties commit, in the context of meaningful mitigation actions and transparency on implementation, to a goal of mobilising jointly USD 100 billion per year by 2020 to address the needs of developing countries
- agreed that, in accordance with paragraph 1(e) of the Bali Action Plan, funds provided to developing country parties may come from a wide variety of sources, public and private, bilateral and multilateral, including alternative sources.

Since then, the issue of measuring, reporting and verifying the USD 100 billion has tended to overshadow the more important and fundamental question of how to mobilise the trillions necessary for the low-carbon transition.

But the scale of investment required is an order of magnitude or more than the Cancun commitment to mobilise USD 100 billion per year by 2020. Much of the finance will need to come from domestic sources and international private investment, underlining the importance of getting the domestic enabling conditions right, including high-quality policy for low-carbon development.

More generally, understanding the impact of ODA on leveraging private investment would enable governments to design appropriate tools and mechanisms to do so. To date, there is very limited understanding of private climate flows to climate change mitigation activities beyond large-scale renewable energy projects, with critical data gaps in transport, water, land-use, energy efficiency and adaptation (Caruso and Jachnik, 2014). Key elements for governments to consider in developing a comprehensive framework for the measurement, reporting and verification of climate change support are: consistent definitions (such as what counts as low-carbon, climate-resilient activities); clear methodologies, (i.e. to estimate private climate finance mobilisation); robust and integrated data management systems; and transparency (Caruso and Jachnik, 2014). They should also build the capacity of individual entities, countries and international data systems for more systematic data collection and reporting (e.g. on private co-financing for publicly supported climate activities).

The OECD-led Research Collaborative on Tracking Private Climate Finance brings together research organisations, international finance institutions and governments. It is investigating and testing options for estimating the mobilisation by developed countries of private finance for climate action in developing countries, as well as ways to improve the underlying availability of data on private climate finance beyond large-scale renewable energy projects.

Notes

1. Throughout this report, "sustainable" is meant as environmentally sound and socially acceptable.
2. The Policy Framework for Investment (PFI) is a comprehensive and systematic approach for improving investment conditions. It helps governments and regions to design and implement policy reforms to create an attractive, robust and competitive environment for domestic and foreign investment.
3. See: www.gov.uk/government/uploads/system/uploads/attachment_data/file/228752/9780108508394.pdf.
4. The United Kingdom's Green Investment Bank is one example (www.greeninvestmentbank.com).

5. The case studies together give a comprehensive view of what may constitute good approaches to successful green procurement, see: www.oecd.org/gov/ethics/best-practices-for-green-procurement.htm.

6. The United Nations Framework Convention on Climate Change (UNFCCC), the Convention on Biological Diversity (CBD) and the United Nations Convention to Combat Desertification (UNCCD), collectively known as the Rio conventions, were established following the 1992 United Nations Conference on Environment and Development in Rio de Janeiro. The developed countries that signed the three Rio conventions in 1992 committed themselves to assist developing countries in implementing them.

7. As recorded in OECD-DAC statistics (OECD-DAC Statistics, 2014).

References

ACCA and Eurosif (2013), "What do investors expect from non-financial reporting?", The Association of Chartered Certified Accountants, London and European Sustainable Investment Forum, Brussels, June, available at: www.accaglobal.com/content/dam/acca/global/PDF-technical/sustainability-reporting/tech-tp-wdir.pdf.

Baron, R. (2014), "The evolution of corporate reporting for integrated performance (RTSD 30)", background paper for the 30th Round Table on Sustainable Development, 25 June 2014, available at: www.oecd.org/sd-roundtable/meetings/30throundtableonsustainabledevelopment.htm.

BNEF (2014), "Clean energy investment falls for second year", Bloomberg New Energy Finance (BNEF) press release, 15 January, available at: http://about.bnef.com/press-releases/clean-energy-investment-falls-for-second-year.

Carbon Tracker Initiative (2014), "Recognising risk, perpetuating uncertainty – A baseline survey of climate disclosures by fossil fuel companies", Carbon Tracker Initiative, available at: www.carbontracker.org/report/climateriskdisclosures.

Carbon Tracker Initiative (2013), Unburnable Carbon 2013: Wasted capital and stranded assets, Carbon Tracker & The Grantham Research Institute, LSE.

Caruso, R. and R. Jachnik (2014), "Exploring potential data sources for estimating private finance", *OECD Environment Working Papers*, No. 69, OECD Publishing, Paris, http://dx.doi.org/10.1787/5jz15qwz4hs1-en.

CDP (2014), "Climate action and profitability: CDP S&P 500 climate change report 2014", report by CDP North America, available at: www.cdp.net/CDPResults/CDP-SP500-leaders-report-2014.pdf.

CDSB (2013), "Corporate reporting in the new age of transparency", Climate Disclosure Standards Board, London, available at: www.cdp.net/en-US/News/Pages/Corporate-reporting-in-the-new-age-of-transparency.aspx.

CDSB (2013), "UK mandatory greenhouse gas reporting", Climate Disclosure Standards Board, London, available at: www.cdsb.net/uk-mandatory-greenhouse-gas-reporting.

Ceres (2014), "Reducing system risks: The Securities and Exchange Commission and climate change", February, Ceres, Boston, Massachussetts, available at: www.ceres.org/files/investor-files/sec-guidance-fact-sheet.

Cochran, I. et al. (2014), "Public financial institutions and the low carbon transition: Five case studies on low carbon infrastructure and project investment", *OECD Environment Working Papers*, No. 72, OECD Publishing, Paris, http://dx.doi.org/10.1787/5jxt3rhpgn9t-en.

Corfee-Morlot, J. et al. (2012), "Towards a green investment policy framework: The case of low-carbon, climate-resilient infrastructure", *OECD Environment Working Papers*, No. 48, OECD Publishing, Paris, http://dx.doi.org/10.1787/5k8zth7s6s6d-en.

CPI (2014), "Moving to a low-carbon economy: The impact of policy pathways on fossil fuel asset values", CPI Energy Transition Series, October, Climate Policy Initiative, available at: http://climatepolicyinitiative.org/publication/moving-to-a-low-carbon-economy.

Dellink, R. et al. (2014), "Consequences of climate change damages for economic growth: A dynamic quantitative assessment", *OECD Economics Department Working Papers*, No. 1135, OECD Publishing, Paris, http://dx.doi.org/10.1787/5jz2bxb8kmf3-en.

Eklin, K. et al. (2014), "Lessons from established and emerging green investment bank models", OECD Green Investment Financing Forum background note, OECD, Paris.

Ernst and Young (2014), "Tomorrow's investment rules – Global survey of institutional investors on non-financial performance", Ernst & Young, available at: www.ey.com/Publication/vwLUAssets/EY-Institutional-Investor-Survey/$FILE/EY-Institutional-Investor-Survey.pdf.

EC (2014), "Non-financial reporting", European Commission, http://ec.europa.eu/finance/accounting/non-financial_reporting/index_en.htm.

EC (2013), *The EU Emissions Trading System (EU ETS)*, European Union Publications Office, Brussels, http://dx.doi.org/10.2834/55480.

FSB (2013), "Global shadow banking monitoring report 2013", Financial Stability Board, Basel, Switzerland, 14 November, available at: www.financialstabilityboard.org/wp-content/uploads/r_131114.pdf.

FSB (2012), "Identifying the effects of regulatory reforms on emerging market and developing economies: A review of potential unintended consequences", Report to the G20 Finance Ministers and Central Bank Governors, Financial Stability Board in co-ordination with the International Monetary Board and the World Bank, available at: www.financialstabilityboard.org/wp-content/uploads/r_120619e.pdf?page_moved=1.

G20/OECD (2014), *G20/OECD High-level Principles on Long-term Investment Financing by Institutional Investors*, available at: www.oecd.org/finance/principles-long-term-investment-financing-institutional-investors.htm.

GCEC (2014), *Better Growth, Better Climate: The New Climate Economy Report*, The Global Commission on the Economy and Climate, Washington, DC, October, available at: http://newclimateeconomy.report/wp-content/uploads/2014/08/GCEC_GlobalReport.pdf.

Greenhouse Gas Protocol (2012), "India GHG program launches with more than 20 leading companies", Greenhouse Gas Protocol, available at: www.ghgprotocol.org/Release-India-GHG-Program-Launch.

GRI (2013), "G4 sustainability reporting guidelines: Reporting principles and standard disclosures", Global Reporting Initiative, Amsterdam, Netherlands, available at: www.globalreporting.org/resourcelibrary/GRIG4-Part1-Reporting-Principles-and-Standard-Disclosures.pdf.

IEA (2014), *World Energy Investment Outlook*, Special Report, OECD/IEA, Paris, available at: www.iea.org/publications/freepublications/publication/WEIO2014.pdf.

IEA (2013), *Redrawing the Energy-Climate Map: World Energy Outlook Special Report*, International Energy Agency, Paris, available at: www.iea.org/publications/freepublications/publication/WEO_RedrawingEnergyClimateMap.pdf.

IIRC (2013), "The international <IR> framework", The International Integrated Reporting Council, available at: www.theiirc.org/wp-content/uploads/2013/12/13-12-08-THE-INTERNATIONAL-IR-FRAMEWORK-2-1.pdf.

India GHG Program (2014), "Responsible Indian businesses track and manage their own greenhouse gas emissions", India GHG Program, Mumbai, India, 2 December, available at: http://indiaghgp.org/release-responsible-indian-businesses-track-and-manage-their-own-greenhouse-gas-emissions#sthash.EjLqSI4W.dpuf (accessed 10 January 2014).

Indian Institute of Corporate Affairs (2013), "National voluntary guidelines: Creating a conducive policy environment for responsible business and responsible investment in India", available at: http://iipcollaborative.org/wp-content/uploads/sites/3/2014/06/NVG-Genesis-Document.pdf.

Kauffmann, C. et al. (2012), "Corporate greenhouse gas emission reporting: A stocktaking of government schemes", *OECD Working Papers on International Investment*, No. 2012/01, OECD Publishing, Paris, http://dx.doi.org/10.1787/5k97g3x674lq-en.

KPMG et al. (2013), "Carrots and sticks: Sustainability reporting policies worldwide, today's best practice, tomorrow's trends (2013 edition)", KPMG Advisory N.V., Global Reporting Initiative, Unit for Corporate Governance in Africa, available at: www.globalreporting.org/resourcelibrary/carrots-and-sticks.pdf.

Miyamoto, K. and K. Biousse (2014), "Official support for private sector participation in developing country infrastructure", *OECD Development Co-operation Working Papers*, No. 19, OECD Publishing, Paris, http://dx.doi.org/10.1787/5jz14cd40nf0-en.

Narbel, P.A. (2013), "The likely impact of Basel III on a bank's appetite for renewable energy financing", *Norwegian School of Economics Department of Business and Management Science Research Discussion Papers*, No. 2013/10, http://dx.doi.org/10.2139/ssrn.2341519.

OECD (2015a), *Policy Guidance for Investment in Clean Energy Infrastructure: Expanding Access to Clean Energy for Green Growth and Development*, OECD Publishing, Paris, http://dx.doi.org/10.1787/9789264212664-en.

OECD (2015b), *Policy Framework for Investment 2015*, OECD Publishing, Paris.

OECD (2015c), *Overcoming Barriers to International Investment in Clean Energy*, Green Finance and Investment, OECD Publishing, Paris, http://dx.doi.org/10.1787/9789264227064-en.

OECD (2015d), *OECD Business and Finance Outlook 2015*, OECD Publishing, Paris.

OECD (2015e), *Recommendation of the Council on Public Procurement*, C(2015)2, OECD, Paris, available at: http://acts.oecd.org/Instruments/ShowInstrumentView.aspx?InstrumentID=320&InstrumentPID=348&Lang=en&Book=False.

OECD (2015f), "Smart procurement: Going green – Best practices for green procurement", GOV/PGC/ETH(2014)1/REV1, OECD, Paris, available at: www.oecd.org/officialdocuments/publicdisplaydocumentpdf/?cote=GOV/PGC/ETH%282014%291/REV1&docLanguage=En.

OECD (2015g), *Mapping Channels to Mobilise Institutional Investment in Sustainable Energy*, Green Finance and Investment, OECD Publishing, Paris, http://dx.doi.org/10.1787/9789264224582-en.

OECD (2014a), "Annual Survey of Large Pension Funds and Public Pension Reserve Funds", OECD, Paris.

OECD (2014b), *Development Co-operation Report 2014: Mobilising Resources for Sustainable Development*, OECD Publishing, Paris, http://dx.doi.org/10.1787/dcr-2014-en.

OECD (2014c), "Estimating mobilised private climate finance: Methodological approaches, options and trade-offs", *Policy Perspective,* OECD Publishing, Paris.

OECD (2014d), "Financing climate change action", *Policy Perspectives*, OECD, Paris, available at: www.scribd.com/doc/239900170/Financing-Climate-Change-2014-Policy-Perspectives.

OECD (2014e), "Measuring and monitoring external development finance", element 11, paper 1, OECD ➔ post 2015, OECD, Paris, available at: www.oecd.org/dac/Monitoring%20and%20measuring%20external%20development%20finance.pdf.

OECD (2014f), "Private financing and government support to promote long-term investments in infrastructure", report to the G20, September, OECD, Paris, available at: www.oecd.org/daf/fin/private-pensions/Private-financing-and-government-support-to-promote-LTI-in-infrastructure.pdf.

OECD (2013a), *Putting Green Growth at the Heart of Development*, OECD Green Growth Studies, OECD Publishing, Paris, http://dx.doi.org/10.1787/9789264181144-en.

OECD (2013b), *Government at a Glance 2013*, OECD Publishing, Paris, http://dx.doi.org/10.1787/gov_glance-2013-en.

OECD (2012a), *OECD Environmental Outlook to 2050: The Consequences of Inaction*, OECD Publishing, Paris, http://dx.doi.org/10.1787/9789264122246-en.

OECD (2012b), "Arrangement on officially supported export credits", TAD/PG(2012)9, OECD, Paris, available at: www.oecd.org/officialdocuments/publicdisplaydocumentpdf/?doclanguage=en&cote=tad/pg(2012)9.OECD (2011), Towards Green Growth, OECD Green Growth Studies, OECD Publishing, Paris, http://dx.doi.org/10.1787/9789264111318-en. OECD (2004), *OECD Principles of Corporate Governance*, OECD, Paris, www.oecd.org/corporate/oecdprinciplesofcorporategovernance.htm.

OECD (2002), *OECD Recommendation of the Council on Improving the Environmental Performance of Public Procurement*, C(2002)3, OECD, Paris, available at: http://acts.oecd.org/Instruments/ShowInstrumentView.aspx?InstrumentID=46&InstrumentPID=43&Lang=en&Book=False.

OECD DAC Statistics (2014), *Climate-related development finance in 2013: Improving the statistical picture*, OECD, November 2014. www.oecd.org/dac/environment-development/Climate-related%20development%20finance%20FINAL.pdf

OECD/ITF (2015), ITF Transport Outlook 2015, OECD Publishing, Paris, http://dx.doi.org/10.1787/9789282107782-en.

Reuters (2014), "China tells firms to start reporting carbon emissions", Reuters, Beijing, 18 March, available at: www.reuters.com/article/2014/03/18/china-carbon-idUSL3N0MF29820140318 (accessed 10 January 2015).

Spencer, T. and J. Stevenson (2013), "EU low carbon investment and new financial sector regulation: What impacts and what policy response?", *IDDRI Working Papers*, No. 05/13, April, Institut du développement durable et des relations internationals, Paris, available at: www.iddri.org/Publications/Collections/Idees-pour-le-debat/WP0513_TS%20JS_financial%20regulation.pdf.

Sustainable Stock Exchanges Initiative (2014), "Sustainable Stock Exchanges 2014 report on progress", prepared for the Sustainable Stock Exchanges 2014 Global Dialogue, available at: www.sseinitiative.org/wp-content/uploads/2012/03/SSE-2014-ROP.pdf.

Sutherland, D. et al. (2009), "Infrastructure investment: Links to growth and the role of public policies", *OECD Economics Department Working Papers*, No. 686, OECD Publishing, Paris, http://dx.doi.org/10.1787/225678178357.

UK Department for Environment, Food and Rural Affairs (Defra) (2013), Environmental Reporting Guidelines: Including mandatory greenhouse gas emissions reporting guidance, Crown copyright, London, June, available at: www.gov.uk/government/uploads/system/uploads/attachment_data/file/206392/pb13944-env-reporting-guidance.pdf.

UNCTAD (2014), *World Investment Report 2014: Investing in the SGGs: An Action Plan*, United Nations Conference on Trade and Development, New York and Geneva, available at: http://unctad.org/en/PublicationsLibrary/wir2014_en.pdf.

UNEP (2015), "Aligning the financial system with sustainable development: Pathways to scale", The Inquiry's 3rd Progress Report, United Nations Environment Programme, Geneva, January, available at: http://apps.unep.org/publications/index.php?option=com_pub&task=download&file=-Aligning_the_financial_system.pdf.

UNEP (2014), “Aligning the financial system with sustainable development: An invitation and background briefing”, UNEP Inquiry into the Design of a Sustainable Financial System, United Nations Environment Programme, Geneva, June, available at: www.unepfi.org/psi/wp-content/uploads/2014/07/Aligning_financial_system_with_sustainable_development.pdf.

US EPA (2014), “Greenhouse Gas Reporting Program. Basic information”, United States Environmental Protection Agency website, www.epa.gov/ghgreporting/basic-info/index.html.

WBCSD (2013), “Reporting matters: Improving the effectiveness of reporting – WBCSD baseline report 2013”, World Business Council on Sustainable Development, Geneva, available at: www.wbcsd.org/Pages/EDocument/EDocumentDetails.aspx?ID=15951&NoSearchContextKey=true.

WEF (2014), “Infrastructure investment policy blueprint”, World Economic Forum, Geneva, February, available at: www3.weforum.org/docs/WEF_II_InfrastructureInvestmentPolicyBlueprint_Report_2014.pdf.

WRI (2014a), “China moves toward mandatory corporate greenhouse gas reporting”, World Resources Institute, Washington, DC, available at: www.wri.org/our-work/top-outcome/china-moves-toward-mandatory-corporate-ghg-reporting (accessed on 10 January 2015).

WRI (2014b), “The trillion dollar question: Tracking public and private investment in transport”, Working Paper, World Resources Institute, Washington, DC, January, available at:www.wri.org/sites/default/files/trillion_dollar_question_working_paper.pdf.

Chapter 3

Implementing climate-friendly taxation practices

Taxation is an important lever of economic policy. Taxes and tax expenditures on energy can greatly influence energy-related CO_2 emissions. After an overview of existing fossil fuel subsidies, this chapter first describes the diversity of taxes applying to fossil fuels, including the differences between gasoline and diesel fuels in transport. It then identifies other tax provisions that can have a strong influence on emissions, such as the fiscal treatment of company cars and commuting expenses, and the design of property taxes. The role of tax provisions in driving investments in specific activities should also be assessed against the objective of the low-carbon transition.

Key messages

Taxation is an important determinant of economic choices. Fiscal policy is also an important element of a country's economic strategy, including social equity. In the wake of the financial crisis and with budget consolidation needs, governments have been looking for new tax revenues, including taxes on greenhouse gases and other environmental externalities.

Are energy-related taxes and tax expenditures conducive to low-carbon choices? One of the obvious misalignments with the low-carbon transition is the existence of subsidies and tax expenditures favouring the production and use of fossil fuels. These instruments are increasingly hard to justify if their goal is to bring more fossil fuel to the market. Where their goals are social, they are often poorly targeted and lead to higher greenhouse gas emissions and other external costs. Fossil fuel subsidies can adversely affect the greenhouse gas implications of international trade, e.g. by distorting markets and harming the competitiveness of renewables and energy efficient technologies. Reform in this area is not easy, but possible: Indonesia has increased the price of diesel by 67% and that of gasoline by 89% through subsidy cuts since 2013. The current trend of low oil prices can facilitate such policy reform, as well as the introduction of higher taxes on oil products where suitable.

Energy taxes, when expressed on a per-tonne-of-CO_2 basis, vary significantly across fuels and end-uses, indicating potential for increased tax revenues and a more homogenous price on CO_2 emissions. Tax differentials between diesel and gasoline, or lower value-added tax rates applied to energy products, continue to encourage higher energy use with negative impacts on the local and global environment. Fuels used for heating and electricity generation are often lightly taxed, if at all.

Are there policy misalignments in tax provisions outside energy? Another misalignment is found in the favourable personal tax treatment of company cars (19% of OECD countries' total car fleet) and commuting expenses in many countries. These tax provisions encourage a higher number of cars that are used more intensively, resulting in increased CO_2 emissions and other external costs.

Property taxes and related instruments can also influence future CO_2 emissions, especially in countries with rapidly growing urban areas. Taxes on property can encourage urban expansion, a problem that can be fixed by changing the tax base. The role of tax provisions favouring home ownership over other household investment is also under scrutiny for its possible effects on employment and mobility needs. As is the case with tax policy more generally, this issue should be cast in light of the broader economic benefits of such tax measures, in country-specific contexts.

Taxes and tax expenditures on corporate income are powerful drivers of economic choices, including investment. A preliminary survey of tax provisions in G20 countries favouring investment in specific activities or regions indicates occasional biases in favour of energy-intensive activities. These could inadvertently enhance the risk of stranding assets and warrant closer examination.

How would the low-carbon transition affect future tax revenues? The low-carbon transition sometimes raises concerns about future tax revenues from fossil fuel use. If a CO_2 tax were introduced to reduce emissions, projections show that its revenues would more than offset the reduction in energy tax revenues for some time to come. Governments nevertheless need to anticipate the impacts of the low-carbon transition on tax revenues. The possible reform of specific tax provisions to better align incentives with the low-carbon transition would need to be integrated into this discussion.

The taxing issue of low-carbon economies

Tax systems influence firms' and individuals' choices. The broad structure of most tax systems emerged at a time when scarcity of natural capital and environmental and health damage were low on the agenda. While taxes were levied on natural resources (e.g. fossil fuel royalties and land taxes), and sometimes on goods with large environmental impacts (e.g. transport fuel), their revenue-raising role was much more prominent than resource management or environmental protection.

An increasing number of countries have made progress in environmental tax reform, loosely defined as the use of market-based instruments – including taxes and tradable permits – to improve the extent to which prices reflect all of the social and environmental costs of economic activities.[1] Using taxes or charges to incorporate these "external" costs leads to a more efficient economy.

This chapter investigates traits of the tax system that will need reform if carbon pricing and other climate policy instruments are to effectively drive the low-carbon transition. However, wider market barriers and failures (see Chapter 1) may mean that environmental tax reform is a necessary (but not always sufficient) condition for efficient use of natural resources, protection of the environment and transition to a low-carbon economy.

Taxes are said to be efficient when they raise revenues without distorting economic decisions. Environmental taxes and the reforms identified in this chapter play a different role: they seek to improve efficiency by bringing prices more in line with marginal social costs and changing economic decisions accordingly. In addition, the productive use of tax revenues or subsidy reform enables governments to reduce other, possibly less efficient, types of taxes. For example, *Tax Policy Reform and Economic Growth* (OECD, 2010) ranks taxes according to their impact on long-run per capita gross domestic product (GDP) as a proxy for well-being: recurrent taxes on immovable property are the least harmful, followed by consumption taxes (including environmental taxes) and other property taxes, personal income taxes and corporate income taxes. An environmental tax reform could therefore reduce the tax system's negative impact on growth if it were combined with reduced reliance on personal and corporate income taxation (Box 3.1).

Box 3.1. A second look at taxing carbon: Beyond budget consolidation

What could be the role of carbon pricing in the context of the broader tax policy reform? In principle, carbon taxes are introduced to restore economic efficiency by reflecting the social cost of carbon in relative prices. The effects on long-run GDP are through reduced damage from climate change.

A new carbon tax would also bring additional revenues. Cournède et al. (2013) assess environmental taxes alongside other possible sources for fiscal consolidation, but find them to be mildly harmful for short-term equity and growth. Flues (2015) shows how the equity impact of carbon and energy taxes can be more than compensated through the recycling of a portion of these taxes' revenues. The growth effect, which does not take the long-term reduction of climate impacts into account, can be mitigated by the reduction of other, more distortive taxes.

The role of a carbon tax should be considered in the context of a broader fiscal reform agenda. Recent modelling work on 35 developed and developing countries finds that carbon tax revenues, recycled through either lump-sum transfers, deficit reduction, labour taxes or government investment, can lower or more than compensate the short-run effects of the carbon tax on growth. Such modelling exercises tend to overestimate negative GDP impacts, as they cannot capture all distortions embedded in the tax system (GCEC, 2014).

The distributional aspects of the low-carbon transition and how taxation can address any negative impacts (see Chapter 1) are essential elements of a successful low-carbon policy strategy. Without a clear and comprehensive strategy for dealing with distributional impacts, policy makers may not be in a position to push low-carbon policy forward.[2]

This chapter first identifies negative features of tax systems for the low-carbon transition. It then investigates to what extent the tax code at large is or is not neutral with respect to low-carbon investment choices.

How energy subsidies and taxes undermine climate change action

This section explores several misalignments between energy taxes and subsidies and the low-carbon transition, looking beyond the lack of explicit CO_2 prices in many jurisdictions.

Tax expenditures and fossil fuel subsidies

A number of tax expenditures[3] encourage the economy to produce or consume more fossil fuels than it would in the absence of such measures. While such an approach may have been justified on grounds of energy independence, it will be increasingly inconsistent with governments' aspirations for climate change mitigation. The OECD's *Inventory of Estimated Budgetary Support and Tax Expenditures for Fossil Fuels* lists a total of 550 measures supporting coal, oil and gas production and use across the 34 OECD countries (OECD, 2013a). Together these were worth USD 55-90 billion every year between 2005 and 2010. As OECD countries remain mostly fossil fuel importers, more than two-thirds of this support goes to the consumption of fossil fuels, sometimes in the form of a price-fixing mechanism, which can become a large subsidy when international prices soar.[4] While not all these expenditures mean lower fossil fuel prices for consumers, they enhance the profitability of activities related to fossil fuels. This generally contradicts the need to direct the economy away from CO_2-emitting fossil fuels, and also affects governments' budgets.

Outside the OECD, several countries also maintain domestic energy prices below international benchmark prices. The International Energy Agency (IEA) estimates that fossil fuel subsidies, based on a price-gap analysis, had reached USD 548 billion in 2013 (IEA, 2014). The IEA lists over 40 countries, representing well over half of global fossil fuel consumption, where at least one fossil fuel is subsidised when compared to its benchmark international price. Most of these subsidies are found in developing countries. Some reform has been undertaken or is underway to reduce and replace them with other social transfers when necessary, but these estimates indicate that more could be done to remove unjustified support for fossil fuel production and use. The impacts of fossil fuel support are indeed quite striking in terms of climate policy costs. For instance, without subsidies to oil use in power generation in the Middle East, oil-fired power plants could not compete with any of the main renewables or nuclear technologies (IEA, 2014).[5]

The piecemeal approach to energy taxes is harmful

Policy instruments to price greenhouse gas emissions (GHGs) are generally applied on top of existing taxes which already affect production and consumption patterns, e.g. different tax rates applied to gasoline and diesel. Many countries' energy tax rates vary considerably for similar fuels across sectors. "Many differentials may, however, have simply arisen out of the piecemeal design and introduction of taxes on different energy products at different points in time" (OECD, 2013b).

These differences result from policy choices made on grounds including resource availability, social justice (limited taxes on fuel oil for residential heating), competitiveness considerations (low or zero taxes on coal use in industry or exemptions for agriculture), local externalities (lower taxes on energy used in mass transport) or fiscal efficiency (high taxes on transport fuels, where price elasticity is low and tax revenues more stable) (OECD, 2013b). The result is that there is no uniform price signal on CO_2 emissions in OECD countries that would be consistent with a social cost of carbon (Figure 3.1). Despite being averaged across OECD countries, the figure shows that the differences are still significant.

The above evidence shows that there are opportunities to lower energy-related CO_2 emissions at a low cost by raising the lowest tax rates or levying taxes on sectors which are untaxed as of now. One should not infer that transport fuels should be "let off the hook" given their high tax rate: the high effective CO_2 price reflects, in fact, many other externalities such as local pollution or the cost of road infrastructure. CO_2 emissions reductions are still required in the transport sector as well, which a price signal could help achieve.

Figure 3.1. **Taxation of energy in the OECD area on a carbon content basis**

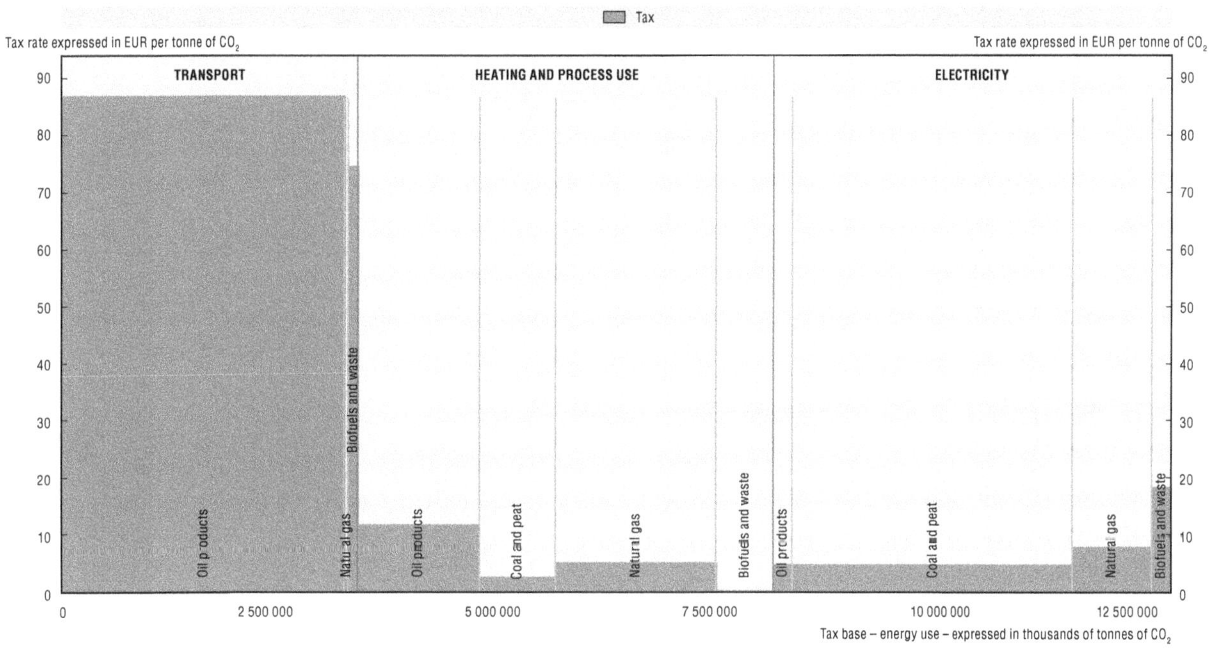

Source: OECD (2013b), *Taxing Energy Use: A Graphical Analysis*, OECD Publishing, Paris, http://dx.doi.org/10.1787/9789264183933-en.

Value added taxes on energy

Value added taxes (VAT) often make up a substantial part of the end-use energy price and, as such, deserve close scrutiny. OECD (2015) considered VAT rates on energy products for OECD countries and Argentina, Brazil, the People's Republic of China (hereafter "China"), India, Indonesia, the Russian Federation and South Africa. It tested whether these rates differ from standard VAT rates – an undifferentiated VAT rate would not change relative prices between energy and other goods. A differentiated VAT rate, however, affects relative prices; a reduced rate, in particular, partly offsets the impact of specific excise taxes on energy products.[6]

Out of the 41 countries studied, 16 have lower VAT rates for energy products: "the reduced rates are either set at approximately half the standard VAT rate or are substantially reduced to rates between 3% and 7%. Reduced or zero VAT rates are most frequently applied to electricity (13 countries), firewood (9), heating oils (9), natural gas (8), liquefied petroleum gas (3), coal (3) and, kerosene or aviation fuel (3), crude oil (3), diesel (2) and gasoline (1). Differential rates may apply only to specific users, e.g. all or some households, or small businesses" (ibid.).

Lower VAT rates tend to be explained politically by the high share of energy expenditures in lower income household budgets, or specific economic activities (agriculture, fisheries). However, higher income households often benefit more from low rates in absolute terms than lower income households. There are more effective means of transferring resources to poor households in a way that does not encourage the use of fossil fuels (see Chapter 1, and Flues, 2015).

The ups and downs of international energy prices

Even if taxes sometimes account for the majority of the final price of energy (e.g. for gasoline or diesel in some OECD countries), end-user prices also reflect the underlying commodity prices, although in a less immediate fashion in markets with price regulation.

The ups and downs of international energy prices clearly affect incentives to reduce GHG emissions, and to invest in exploration and production of marginal oil and gas fields. A good example is the current case of car users experiencing a drop in gasoline prices. All things being equal, the lower prices are likely to lead to greater car use, and therefore more CO_2 emissions. It also means people are less inclined to buy more efficient vehicles, locking in high emissions for some time. International prices also affect the cost of low-carbon policies. In the case of electricity plants capped under an emissions trading system, a drop in coal prices would encourage more coal use. For a given emissions cap, this automatically triggers an increase in the price of CO_2 allowances paid by sectors beyond electricity (Braathen, 2011). International energy price movements therefore have a bearing on the effectiveness of low-carbon policies and their cost.

A drop in international energy prices raises the possibility of increasing taxes or removing subsidies on fossil fuels, with less impact on the incomes of those benefiting from the subsidies. Tax revenues can be redistributed through cuts in other distortive taxes used for budget consolidation or earmarked for low-GHG technologies. Policy makers could flag to the general public that the increase in tax is to: 1) reduce other distortive taxes or increase other transfers; 2) keep emissions under control as part of climate policy. It is important, however, to keep this arrangement stable over time, especially from a budget perspective. The new tax should therefore not be adjusted downward if international prices go back up.

Generally speaking, countries that are keen on taxing carbon should prepare for situations where the price falls on international markets, inadvertently undermining their efforts to reduce CO_2 emissions and other externalities – recognising, at the same time, the advantages of low-energy import prices on trade balances and households' available income. There can be both a carbon and a growth dividend during such episodes, provided policies are in place and the general public understands the strategy underlying the carbon tax.

The diesel-gasoline tax differential makes little sense

Of the 34 OECD countries, 30 have lower tax rates on diesel than on gasoline (Harding, 2014a). This tax differential is at odds with the environmental effects of each fuel, including CO_2 emissions. A litre of diesel emits about 18% more CO_2 than a litre of gasoline. All OECD countries, with the exception of the United States, tax diesel less (and often much less) than gasoline when measured on a per unit of energy basis or a tonne of CO_2 basis. This differential is of course even more pronounced in countries with a tax preference for diesel (Figure 3.2).

Figure 3.2. **The difference between gasoline and diesel tax rates for road transport**

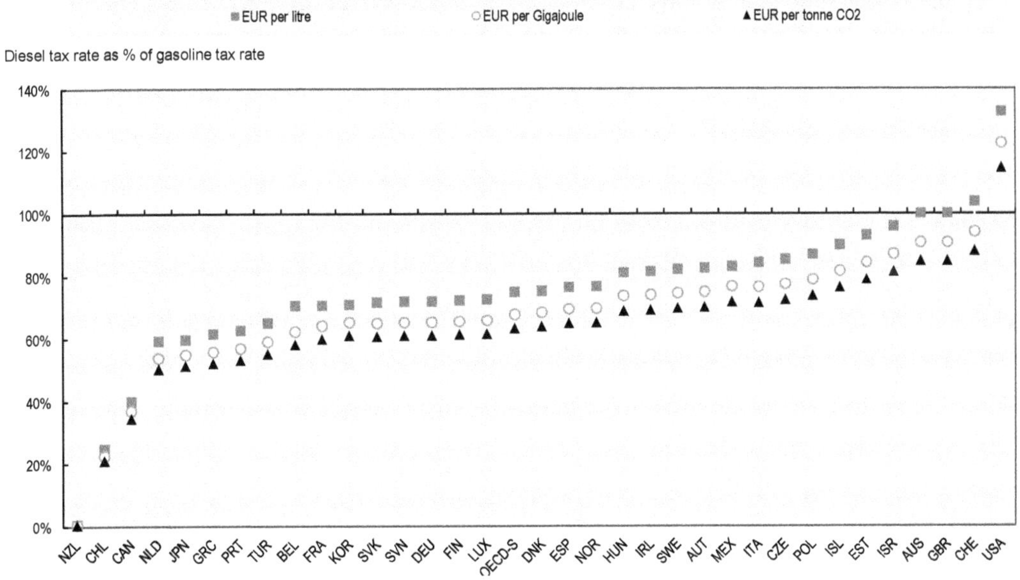

Notes: France reduced the gasoline-diesel tax differential by EUR 0.02 from its initial level of EUR 0.18. New Zealand levies a per kilometre road user charge, with higher charges for larger vehicles; although the charges are not on fuel *per se*, they essentially equalise the costs of diesel and petrol in New Zealand. Tax rates are as of 1 April 2012 (except for Australia for which it is 1 July 2012). Figures for Canada and the United States include only federal taxes. OECD-S is the simple OECD average; OECD-W is the weighted OECD average.

Source: Harding, M. (2014a), "The diesel differential: Differences in the tax treatment of gasoline and diesel for road use", *OECD Taxation Working Papers*, No. 21, OECD Publishing, Paris, http://dx.doi.org/10.1787/5jz14cd7hk6b-en.

The tax differential tends to be justified as a support to users of diesel in the agriculture, fisheries and road freight industries, which also compete internationally, making it more difficult to reduce the differential. It is sometimes justified by the belief that greater efficiency of the diesel engine would mean fewer CO_2 emissions overall. However, the efficiency advantage does not warrant a lower tax because the environmental and social costs are, in fact, higher per unit of energy for diesel than for gasoline (especially given diesel's higher nitrous oxide and particulate emissions; OECD, 2014). The higher fuel efficiency of diesel cars may mean lower CO_2 emissions per kilometre travelled, but only if the diesel vehicle is at least 15% more efficient than the gasoline vehicle it displaces, which is not always the case (Harding, 2014a). Moreover, the economic benefit of better fuel economy of diesel vehicles is entirely captured by their owners – not society at large – meaning that no subsidy is needed. A lower diesel price may encourage the greater use and purchase of larger diesel vehicles, resulting in more CO_2 emissions.

The tax differential on diesel is not justified on environmental nor other grounds. Its gradual removal would mean a cleaner environment and fewer CO_2 emissions. Instead, the tax rate would need to include a CO_2 component per litre, i.e. a higher rate on diesel per litre. In parallel, the impact of existing purchase or registration taxes on vehicle choice, vehicle characteristics and driving behaviour should be carefully considered alongside the levels of fuel tax in each country (ibid.).

Beyond energy taxes: Tax signals hindering low-carbon choices

Tax treatment of company car use and commuting expenses

Company cars represent a substantial share of the car stock in many OECD countries.[7] Tax regimes on company car use and commuting expenses can favour certain modes of transport over others and influence how much employees travel; in addition, few countries tax the benefit of free parking provided by employers. The implications for CO_2 emissions are clear (Harding, 2014b).

The carbon impact of company car tax regimes depends on incentives provided to employers and employees to increase the number of cars on the road, their fuel efficiency and distances driven. Under-taxation of the capital cost of a car may increase the number of cars in a country. The favourable tax treatment of fuel expenditures discourages the purchase of more efficient company vehicles. And the distance travelled is likely to be higher where the tax treatment of the private use of company cars is favourable.

One study finds considerable aggregate taxable benefits for company cars across 27 OECD countries and South Africa (ibid.). Based on a benchmark tax treatment, the study estimates that the weighted average subsidy per company car per year is EUR 1 600. The total value of the tax expenditures related to company cars and their use across countries studied ranges from EUR 19.0 billion to EUR 33.7 billion, with a midpoint estimate of EUR 26.8 billion.

As the taxable benefits measured are largely insensitive to distance driven, these tax regimes encourage individuals to increase the distance driven in company cars. Moreover, they tend to provide a greater subsidy to cars that are less fuel efficient, leading to higher associated CO_2 emissions than would be the case with a neutral tax treatment. Company cars also tend to be larger and less efficient than private cars (ibid.).

The signal sent by tax breaks for company cars tends to contradict the aim of reducing CO_2 emissions and other external costs of transport. Governments could tip the balance in the other direction – towards environmentally motivated company car taxation – to encourage energy efficient vehicles. However, the efficiency of such measures should be assessed against that of other instruments and policies more closely related to CO_2 emissions, such as fuel taxes.

Corporate income taxes: Carbon neutral?

To date, the implications of countries' corporate tax codes for the low-carbon transition remain uncharted territory, with a few exceptions (Box 3.2).

An initial survey of corporate tax provisions in G20 countries, which was conducted by PricewaterhouseCoopers LLP for the OECD and for this project, identifies tax provisions that governments use in support of specific, carbon-intensive activities. These vary in nature:

- Accelerated depreciation of capital: assets are depreciated more rapidly for tax than for accounting purposes, with a benefit for capital-intensive activities, which can be more energy-intensive than average activities. For instance, Australia has tax depreciation rates that may exceed accounting depreciation rates for assets such as aircraft, buses, trucks, tractors and harvesters, as well as some oil and gas-related activities.[8] Spain also has accelerated depreciation on mining assets.
- Other investment incentives and tax holidays are also used: Hungary, Israel and South Africa have investment incentives that are available to all industries but which tend to be used by heavy industry. Indonesia has inbound investment incentives available to specific industries, many of which are GHG-intensive. Some countries provide incentives for investments in economic development or less-developed zones; although they are not specific to an activity, they tend to benefit capital-intensive heavy industry.

Box 3.2. Accelerated depreciation of capital: Insights from a US National Academy of Sciences study

One of the major tax expenditures is the possibility to depreciate equipment and machinery faster than the economic life of these assets would suggest. A recent US study by the Committee of the Effects of Provisions in the Internal Revenue Code on Greenhouse Gas Emissions indicates the important effect of accelerated depreciation on the economy and greenhouse gas emissions (Nordhaus et al., 2013). Accelerated depreciation can influence emissions in the following ways:

- "Positive effects of accelerated depreciation on economic growth could raise GHG emissions, simply because a larger economy, with all other things unchanged, will require greater use of fossil fuels.
- A more capital-intensive economy could raise or lower GHG emissions per unit of output, depending on whether capital and energy are substitutes or complements in production.
- Shifts in output among industries (from those that use less to those that use more machinery and equipment) could increase or decrease fossil fuel use, thereby raising or lowering GHG emissions per unit of output.
- Accelerated replacement of old capital could reduce GHG emissions per unit of output if new capital is more energy efficient than the capital it replaces." (ibid.)

A simulation of the removal of accelerated depreciation finds a small downward impact on the GHG intensity of the US economy, with a low impact on overall emissions as the impact on national output is slightly positive. While the study concludes that the effect may be too small to matter at that resolution of economic model, it seems nonetheless worthwhile to explore how common such tax expenditures are in other countries and how they influence various activities, especially the high capital-intensity of low-carbon power generation technologies.

Further, the interaction between tax provisions and climate objectives ought to be tested in a low-carbon policy, not in a business-as-usual scenario, as was done in the cited study.

Source: Nordhaus, W.D. et al. (eds.) (2013), *Effects of U.S. Tax Policy on Greenhouse Gas Emissions*, Committee of the Effects of Provisions in the Internal Revenue Code on Greenhouse Gas Emissions, Board of Science, Technology, and Economic Policy, Policy and Global Affairs, National Research Council of the National Academies, The National Academies Press, Washington, DC, available at: http://sites.nationalacademies.org/PGA/PGA_084888.

However, there are instances of the use of tax provisions to promote low-carbon or high-efficiency investments (in Ireland,[9] Japan, Mexico and the United Kingdom, among others). Some countries, such as Canada, have taken steps to realign tax depreciation rates for the mining and energy sectors with those for other industries. In the European Union, state aid guidelines do not authorise development tax incentives for activities such as ship-building, steel, coal or investment in energy supply infrastructure.

The above list is not conclusive evidence of widespread corporate tax misalignments with climate policy objectives. These measures are, however, powerful drivers of investment and economic activity – and indeed are sometimes part of a country's industrial policy. Their differentiated use across activities could therefore be evaluated from a climate policy perspective. Any assessment should consider the combined effects of all corporate tax signals. For instance, a mining company may be granted accelerated capital depreciation but pay high levies and royalties on its output.

Property and land taxes need to be better designed

Property taxes, if properly designed, can encourage denser cities that contribute to less private car use (see Chapter 8). A study of the United States finds that a higher property tax reduces city size, all other things being equal (Brandt, 2014). The devil is in the details, however: taxes on land and especially on the value of buildings can hinder renovation and have been said to be the cause of inner city deterioration and urban sprawl (Land, 1967; Carey, 1976; Wyatt, 1994, cited in Brandt 2014). Property taxes are also sometimes designed to encourage single-occupancy housing; this should be avoided if denser housing infrastructures are desired (Merk et al., 2012). This goal is best achieved by combining a two-tier property tax based more on the value of land than on buildings, with low taxes on renovation (Brandt, 2014).[10]

A (higher) property tax could also allow the reduction of other more distortive taxes and make the overall tax system more efficient. In China, sub-national governments raise only a small amount of revenue from property taxes (and more from property sales taxes). They often resort to land conversion fees to balance their budgets. The resulting conversion of agricultural land into urban land, then sold at higher price to developers, generates urban sprawl (Van Dender, 2015). More recurrent taxes on immovable property may be better both for large Chinese cities and the country as a whole (Brys et al., 2013).

France recently introduced an insufficient density payment (Avner et al., 2013). The system does not yet cover the whole country, but can be applied by local governments that have developed urban mobility plans (see Chapter 8). The goal is to limit urban expansion by increasing the construction cost of buildings with lower built-to-land area ratio. This tax on land value has a number of attractive features: it does not affect already built areas, only new developments; it can encourage higher density and more built-floor area, in so doing lowering average rents; and it could also raise significant tax revenues. To achieve these benefits, the minimum threshold has to be carefully calibrated, and the tax should apply across neighbouring jurisdictions.

Rebalancing tax revenues to make greater use of property tax would bring other benefits. In general, investment in housing is under-taxed compared with other investments and savings instruments that would yield higher returns. The under-taxing of housing may also have contributed to housing price bubbles. Another advantage of property taxes is the stability of their tax base, which means more predictable revenues for governments (Brys et al., 2013).

To be efficient, however, property taxes must be regularly updated to reflect the market valuation of real estate: "In many countries cadastral values have become outdated, often by a large margin (by way of example, Austria, Belgium and France last carried out a housing valuation exercise three or four decades ago)" (Cournède et al., 2013). Updating registered property values will also make property taxes less regressive and promote inclusive growth.

Property taxes can also be effective instruments for encouraging low-carbon investment by homeowners. For example, property tax rebates can be offered to encourage energy efficiency or renewable energy investment (Brandt, 2014). However, evidence of the effectiveness of such measures is scant or ambiguous, as they have only been introduced recently. In the United States, the Property Assessed Clean Energy programme (PACE) adds a charge on the property tax that is used to reimburse energy efficiency and renewable energy improvements. This links the renovation investment to the property rather than its occupant and allows for a longer period to repay the investment. Essential steps in designing such taxes involve: 1) checking for any overlaps with other incentives and policies to decarbonise the buildings sector; 2) assessing if they could undermine the other positive effects of the property tax on housing density, etc. highlighted above.

Taxes on property transactions can have a negative, indirect effect on the environment by encouraging the purchase of undeveloped land for new housing far from city centres and infrastructures. Taxes on property transactions also hinder the relocation of homeowners and job mobility, with negative effects on overall productivity (OECD, 2010), leading some to argue for the property sales tax to be abolished (Brandt, 2014). The link between these taxes and reduced workforce flexibility may increase transport use and associated emissions. Statistical analysis across all OECD countries finds that "higher transaction costs in buying a property are associated with lower residential mobility [...] transaction taxes, notarial and legal fees have a negative and significant effect on mobility" (Caldera Sánchez and Andrews, 2011).

The favourable tax treatment of owner occupation over rental in almost all OECD countries may also have knock-on effects on car use in some cases. A study of the impact of home ownership on unemployment rates in the United States finds statistical evidence that "rises in home ownership lead to three problems: *i)* lower levels of labour mobility; *ii) greater commuting times*; and *iii)* fewer new businesses" (Blanchflower and Oswald, 2013; our emphasis). Evidence is needed from other countries to check whether similar mechanisms are also at play or if home ownership systematically brings these important side effects. In addition, home ownership can also have positive effects, as it encourages home energy efficiency improvements, in contrast with landlord-tenant situations.

Where next for tax revenues and budgets in the context of lower fossil energy use?

The low-carbon transition will lead to lower fossil fuel use over the course of the 21st century. In the long run, this may have adverse effects on countries endowed with these resources, which are a source of economic dividends, tax revenues and royalties. The recent fall in international oil prices reveals the significant dependence of these countries on such revenues, including for balancing budgets. The strategic question for these economies is the design of a policy package and broad-based incentives to diversify their structure away from fossil fuels, if they are to cope with the low-carbon transition. Tax changes will not be enough.

In other countries, tax and budget experts worry that environmental taxes will not ensure stable budgetary resources, as they are meant to reduce their own base – unlike property taxes or taxes on profits. In reality, the effect of any new tax on energy consumption is likely to be progressive. In the near term, the instability of energy-related revenues comes mostly from economic cycles, which affect other tax revenues as well.

Further, this issue is of lesser importance, as carbon and other environmentally related taxes do not account for a large share of government revenues in OECD countries (equivalent to 1.57% or 2.25% of GDP in 2013, on a weighted average and arithmetic average basis, respectively).[11] With much lower CO_2 emissions anticipated in the second half of the century, and therefore lower fossil fuel use, the question will gain importance.

For countries that decide to rely on a CO_2 tax as the main price signal to reduce emissions from fossil fuels, the overall impact on tax revenues would be as follows:

- revenues from energy taxes would fall as the use of CO_2-emitting fuels decline
- revenues from the CO_2 tax would increase because the carbon price is projected to rise faster than the reduction rate of emissions.

The scenarios from the OECD model (known as ENV-Linkages) provide useful insights into this issue over the period to 2050. They indicate that revenues from GHG taxes would more than offset the decline in energy taxes as fossil fuel energy demand falls in a scenario compatible with the 2°C Scenario.[12] Obviously, if no taxes are raised on CO_2 and other greenhouse gases – i.e. if other policy instruments are used to achieve the same environmental outcome – governments would need to raise other taxes to compensate for the fall in fossil fuel-energy tax revenues. New energy sources may be taxed at that point.

There may be other competing uses for carbon tax revenues, such as compensating poor households for increased energy prices, recycling them into lower more distortive taxes such as taxes on labour, funding the low-carbon transition with carbon tax revenues or other subjects given priority in ordinary general budget consolidations. All these considerations should be included in projections of future tax revenues and expenditures under a low-carbon scenario. These discussions should also take into account the other tax measures that countries may reform for a more inclusive and pro-growth tax system.

Notes

1. See the recent example of the Green Tax Reform adopted by Portugal in the context of its strategy to come out of the economic and financial crisis.

2. Further, recent OECD work on inequality has shown its direct effects on economic development (Cingano, 2014).

3. A tax expenditure broadly refers to a tax exemption, deduction or credit for selected groups or specific activities. It is a relative preference within a country's tax system that is measured with reference to a benchmark tax treatment set by that country. A tax expenditure implies foregone tax revenues (OECD, 2013a).

4. This was the case in Mexico in 2008. As a result of Mexico's 2013 Tax Reform, the country established, since 2014, a positive tax on hydrocarbons and their derivatives, based on the fuels' carbon content, regardless of the oil price level. With this mechanism, the "negative tax" on fuels has been effectively eliminated.

5. Note that this analysis predated the 2014-15 drop in international oil prices.

6. There is no agreement on whether different VAT rates on energy sources would constitute a subsidy; there is also no internationally agreed definition of fossil fuel subsidies.

7. The median share of company cars in total number of registered cars in 2009-11 was 32% for 26 countries (19% when measured as a weighted average). Sweden (48%), Austria (46%), Hungary (42%) and Belgium (41%) had the largest shares, but these fleets perform relatively well in terms of CO_2 ratings. In terms of the total number of company cars, France, the United Kingdom, Germany and the United States stand out.

8. In Italy, solar and wind-powered electricity generation assets have a tax depreciation rate of 4%, whereas fossil fuel-based generation assets are depreciated at 9%. This is not an explicit decision, but results from wind and solar-powered installations being classified as immovable rather than movable assets.

9. In Ireland, the Accelerated Capital Allowances for Energy Efficient Equipment allows companies to deduct 100% of expenditures incurred on eligible equipment in computing taxable trading profits in the year of purchase rather than over the usual eight-year period for plant and machinery.

10. Various models for property taxation are being discussed for Germany's planned reform of property taxes.

11. Preliminary numbers for 31 OECD countries, not including Australia and Greece, are available in the OECD database on instruments used for environmental policy.

12. ENV-Linkages modelling team, based on *OECD Environmental Outlook 2012* scenarios.

References

Avner, P. et al. (2013), "Modélisation de l'effet d'une taxe sur la construction – Le versement pour sous-densité", *Revue de l'OFCE/Débats et politiques*, No. 128, available at: www.ofce.sciences-po.fr/pdf/revue/14-128.pdf.

Blanchflower, D.G. and A.J. Oswald (2013), "Does high home-ownership impair the labor market?", *National Bureau of Economic Research Working Papers*, No. 19079, May, http://dx.doi.org/10.3386/w19079.

Braathen, NA. (2011), "Interactions between emission trading systems and other overlapping policy instruments", *OECD Green Growth Papers*, No. 2011/02, OECD Publishing, Paris, http://dx.doi.org/10.1787/5k97gk44c6vf-en.

Brandt, N. (2014), "Greening the property tax", *OECD Working Papers on Fiscal Federalism*, No. 17, OECD Publishing, Paris, http://dx.doi.org/10.1787/5jz5pzw9mwzn-en.

Brys, B. et al. (2013), "Tax policy and tax reform in the People's Republic of China", *OECD Taxation Working Papers*, No. 18, OECD Publishing, Paris, http://dx.doi.org/10.1787/5k40l4dlmnzw-en.

Caldera Sánchez, A. and D. Andrews (2011), "To move or not to move: What drives residential mobility rates in the OECD?", *OECD Economics Department Working Papers*, No. 846, OECD Publishing, Paris, http://dx.doi.org/10.1787/5kghtc7kzx21-en.

Carey, G.W. (1976), "Land tenure, speculation, and the state of the aging metropolis", *The Geographical Review*, Vol. 66, No. 3, pp. 253-265, July, www.jstor.org/stable/213884.

Cingano, F. (2014), "Trends in income inequality and its impact on economic growth", *OECD Social, Employment and Migration Working Papers*, No. 163, OECD Publishing, Paris, http://dx.doi.org/10.1787/5jxrjncwxv6j-en.

Cournède, B. et al. (2013), "Choosing fiscal consolidation instruments compatible with growth and equity", *OECD Economic Policy Papers*, No. 7, OECD Publishing, Paris, http://dx.doi.org/10.1787/5k43nxq6dzd4-en.

Flues, F. (2015), "The distributional effects of energy taxes", *OECD Taxation Working Papers*, OECD, Paris, forthcoming.

GCEC (2014), "Chapter 5: Economics of change", in: *The New Climate Economy Report*, Global Commission on the Economy and Climate, available at: http://newclimateeconomy.report.

Harding, M. (2014a), "The diesel differential: Differences in the tax treatment of gasoline and diesel for road use", *OECD Taxation Working Papers*, No. 21, OECD Publishing, Paris, http://dx.doi.org/10.1787/5jz14cd7hk6b-en.

Harding, M. (2014b), "Personal tax treatment of company cars and commuting expenses: Estimating the fiscal and environmental costs", *OECD Taxation Working Papers*, No. 20, OECD Publishing, Paris, http://dx.doi.org/10.1787/5jz14cg1s7vl-en.

Heindl, P. and A. Löschel (2014), "Energy sector reform and its impact on households", Issue Note, OECD Green Growth and Sustainable Development Forum, 13-14 November 2014, OECD, Paris, available at: www.oecd.org/greengrowth/Issue%20Note%20Session%20One%20GGSD%20Forum.pdf.

IEA (2014), *World Energy Outlook 2014*, International Energy Agency, Paris, http://dx.doi.org/10.1787/weo-2014-en.

IEA (2013), *World Energy Outlook 2013*, International Energy Agency, Paris, http://dx.doi.org/10.1787/weo-2013-en.

Land, A.E. (1967), "Toward optimal land use: Property tax policy and land use planning", *California Law Review*, Vol. 55, No. 3, Article 9, available at: http://scholarship.law.berkeley.edu/californialawreview/vol55/iss3/9.

Merk, O. et al. (2012), "Financing green urban infrastructure", *OECD Regional Development Working Papers*, No. 2012/10, OECD Publishing, Paris, http://dx.doi.org/10.1787/5k92p0c6j6r0-en.

Nordhaus, W.D. et al. (eds.) (2013), *Effects of U.S. Tax Policy on Greenhouse Gas Emissions*, Committee of the Effects of Provisions in the Internal Revenue Code on Greenhouse Gas Emissions, Board of Science, Technology, and Economic Policy, Policy and Global Affairs, National Research Council of the National Academies, The National Academies Press, Washington, DC, available at: http://sites.nationalacademies.org/PGA/PGA_084888.

OECD (2015), *Taxing Energy Use 2015: OECD and Selected Partner Economies*, OECD Publishing, Paris, http://dx.doi.org/10.1787/9789264232334-en.

OECD (2014), *The Cost of Air Pollution: Health Impacts of Road Transport*, OECD Publishing, Paris, http://dx.doi.org/10.1787/9789264210448-en.

OECD (2013a), *Inventory of Estimated Budgetary Support and Tax Expenditures for Fossil Fuels 2013*, OECD Publishing, Paris, http://dx.doi.org/10.1787/9789264187610-en.

OECD (2013b), *Taxing Energy Use: A Graphical Analysis*, OECD Publishing, Paris, http://dx.doi.org/10.1787/9789264183933-en.

OECD (2012), *OECD Environmental Outlook to 2050: The Consequences of Inaction*, OECD Publishing, Paris, http://dx.doi.org/10.1787/9789264122246-en.

OECD (2010), *Tax Policy Reform and Economic Growth*, OECD Publishing, Paris, http://dx.doi.org/10.1787/9789264091085-en.

Van Dender, K. (2015), "Environmentally related taxes and sprawl: A review of research findings", paper presented at the Joint Meetings of Tax and Environment Experts, 6 June 2014, COM/ENV/EPOC/CTPA/CFA(2014)23, OECD, Paris, forthcoming.

Wyatt, M.D. (1994), "A critical review of land value taxation as a progressive strategy for urban revitalization, rational land use and tax relief", *Review of Radical Political Economics*, Vol. 26, No. 1, pp. 1-25, March, http://dx.doi.org/10.1177/048661349402600101.

Chapter 4

Delivering innovation and skills for the low-carbon transition

There is a wide range of policy instruments that drive innovation. Strong climate policies are essential to pull innovation in the right direction, but other instruments play a role, and may inadvertently hamper change. This chapter describes the trends in public research, development, and deployment (RD&D), the incentives for private RD&D and innovations, as well as the labour and capital attractiveness of innovative firms in different policy settings. It also touches on whether countries have the right set of skills for the low-carbon transition. Last, specific industries may face regulatory hurdles to innovate to lower their emissions, as illustrated by a case study on cement manufacturing.

Key messages

Policy measures to spur innovation include public investment in basic research, various support measures to encourage private investment in applied research, development and demonstration (RD&D), procurement, protection of intellectual property and support for public-private co-operation, to name just a few.

A strong policy signal to reduce greenhouse gas (GHG) emissions is essential to create the market pull eventually needed to deploy low-carbon innovations. There is evidence that environmental policy can result in higher productivity through innovation, but that certain policy designs can have the opposite effect and hinder competition by favouring incumbents. This is one area of potentially fruitful policy alignment.

Are public RD&D expenditures measuring up against the low-carbon transition challenge? In light of the role of the energy sector in the low-carbon transition, the declining share of publicly funded energy-related research in the International Energy Agency's country RD&D budgets appears problematic – even if other research domains such as information and communication technologies or nanotechnologies can also lead to low-GHG innovations.

Are innovation incentives conducive to competition by new entrants? Countries increasingly use tax incentives to leverage private research and development (R&D). To encourage new entrants and challenge incumbents, tax incentives should be designed so that young and not-yet-profitable companies are eligible for these benefits. This is not yet the case in all countries. Carry-over provisions can fix this problem. Direct support for R&D through grants and awards can be an important complement. The more general policy context also drives the growth performance of young firms. Reforming regulations that inhibit competition or create barriers to exit can be beneficial to these firms.

Could public procurement policy be a better driver of innovation? Demand-side measures can also spur climate-friendly innovation. Public procurement represents around 13% of gross domestic product in OECD countries, though little of this addresses innovation. Certain policy failures in public expenditure management can also prejudice climate-friendly innovation, such as split responsibilities for capital and operating costs.

An important alignment challenge in the area of low-carbon innovation arises from the fact that effective demand measures such as standard-setting are sometimes defined at the national level. International co-ordination in this area would align signals for businesses on a broader basis.

Are labour markets, education and training systems able to address skills gaps for the low-carbon transition? Efforts may be needed to ensure the adoption of innovations by consumers and businesses. The question arises whether the right skills are available both to generate technological change for the transition and to serve the new markets spurred by climate-related policy measures. There is evidence that significant skills gaps exist. A few countries have established monitoring systems to evaluate skills needs related to the low-carbon economy in the broader context of their employment policy.

Misalignments with low-carbon innovation are also found in regulations affecting inputs, products and services. Innovation in resource efficiency and potential CO_2 reductions is often hindered by regulations that have not caught up with technology. Much effort will be required to inventory these regulatory hurdles and revisit them in light of the carbon constraint.

Unpacking innovation policy for the low-carbon transition

The interplay between policies and innovation in the context of climate policy is multi-dimensional. It is about the creation of new businesses and the end of old ones, the emergence of nascent technologies and business models, and how the policy and regulatory environment must adapt to let them thrive. Clear and credible government decisions to move forward with ambitious core climate policy instruments is an important spur for low-carbon innovation within companies and sectors that are directly targeted by such policies, but also by their suppliers and new entrants, whose products may have a competitive advantage in a low-carbon economy. Overall, low-carbon innovation will also be driven by robust economic signals to business, beyond innovation policy strictly speaking.

It is important to emphasise that direct government support to low-GHG technological change can make economic sense in combination with GHG-pricing policy (Acemoglu et al., 2012; see also OECD, 2012a for a discussion). In effect, two failures need to be addressed simultaneously. On the one hand, carbon emissions generate negative environmental externalities. On the other hand, innovation is a positive externality, as elements of the information and knowledge generated by innovation are public goods.

In general, different instruments should be used to address these two distinct market failures. Therefore, a two-pronged strategy is motivated by the existence of market failures and barriers which constrain both invention and adoption of new technologies. Policy measures to spur innovation include public investment in basic research, various support measures to encourage private investment in applied R&D, protection of intellectual property, support for public-private co-operation and a host of other measures.

In the context of this report the issue is whether the types of policies that governments have introduced to address the latter (i.e. investment in basic research, intellectual property rights, tax incentives and direct support for business R&D, public-private co-operation and networks) are aligned or misaligned with efforts to bring about a transition to a low-carbon economy. More indirect policy conditions, such as those which support or hinder the entry and growth of young and innovative firms, may play an important indirect role. Skills are also a critical link in the transition and deserve policy attention as well.

As case study material, this chapter presents another example of policy misalignment, related to regulations applied to specific industries, which could hinder the uptake of new industrial practices that would lead to greater efficiency of material conversion and energy use, and lower GHG emissions. These inadvertent regulatory barriers warrant more in-depth analysis across all sectors of importance in the low-carbon transition, which is beyond the scope of the present report.

Public investment in research

The financial crisis did not spare innovation policy. Since 2008, OECD countries' gross public expenditures on R&D have been growing at half the annual rate observed in 2001-08. Public R&D budgets have levelled off or started to decline with some exceptions (e.g. Germany and Sweden). Business expenditures on R&D have been growing since 2008 at a pace that is about a quarter of what it was in 2002-08.[1] Nonetheless, investment in knowledge-based capital overall (i.e. including other intangibles) was more resilient than investment in physical capital (OECD, 2015c).

However, it is important to distinguish between basic and applied research – although the distinction is often more a question of degree and not kind – and the important role that the public sector plays in the former. Public research plays a key role in innovation systems by providing new knowledge and pushing the knowledge frontier. Universities and public research institutions (PRIs) often undertake longer term, higher risk research and complement the research activities of the private sector. Although the volume of public R&D is less than 30% of total OECD R&D (OECD, 2014a), universities and PRIs perform more than three-quarters of total basic research. Ensuring a balance between pure basic research, driven by excellence, and more focused, mission-oriented research is therefore an important challenge for public funding. One potential source of misalignment in the context of climate-related innovation is the greater focus on applied research with shorter term objectives (see OECD, forthcoming).

Another important phenomenon is the growth of public R&D activities in emerging economies: non-OECD countries have seen their share in global R&D rise from 10% to 30% of the global total within ten years (OECD, 2014a). As emerging economies make an effort to switch to higher value-added activities, with a parallel R&D effort, it will be important to ensure that this occurs in a way that supports rather than hinders the low-carbon transition.

The case of climate change also highlights the need for new interdisciplinary and trans-disciplinary research environments. The diverse sources of knowledge upon which "climate" innovation, and environmental innovation more generally, draws makes this particularly important in the context of policy alignment for the low-carbon transition (see Chapter 4 of the *OECD Innovation Strategy*, forthcoming). Nanotechnology is a case in point (OECD, 2013a). While the initial spur for research on nanotechnology was not related to environmental considerations, "green" nanotechnology is increasingly being applied in fields such as green chemistry, sustainable manufacturing, and monitoring and control applications (e.g. nano-sensors).

The role of information and communication technologies (ICTs) is another case in point. Decentralised production of energy, which has seen a massive surge in the last few decades, has become possible because of ICTs. Other applications of ICT that can yield environmental benefits include transport logistics and environmental monitoring. For example, with respect to the latter, satellite technologies are essential to the monitoring of atmospheric (e.g. pollution concentrations, wind speed, temperature), oceanic (e.g. sea salinity, level, temperature) and terrestrial (e.g. glaciers and ice caps, ice sheets, albedo, vegetation type, soil moisture) data.[2]

Bearing in mind the wide breadth of research domains which can yield benefits in terms of climate mitigation and adaptation, it is instructive to examine trends in public spending RD&D in the energy sector. While the share of RD&D has fallen from over 10% to less than 5% in the last four decades, the composition has changed. In particular, the past few years have witnessed a remarkable increase in support to renewable energy sources and energy efficiency (Figure 4.1). The surge in fossil fuel-related expenditures in 2009 is explained by US spending on coal, including high-efficiency combustion technology. Overall, in an earlier assessment, the IEA (2013) indicated that global government spending on energy RD&D should triple to match countries' aspirations for low-carbon technologies in particular. A detailed country-level analysis for IEA countries also revealed "some significant discrepancies between stated energy RD&D priorities and actual funding" (IEA, 2013: 129).

Figure 4.1. **Public sector spending in energy research, development and demonstration in IEA countries**

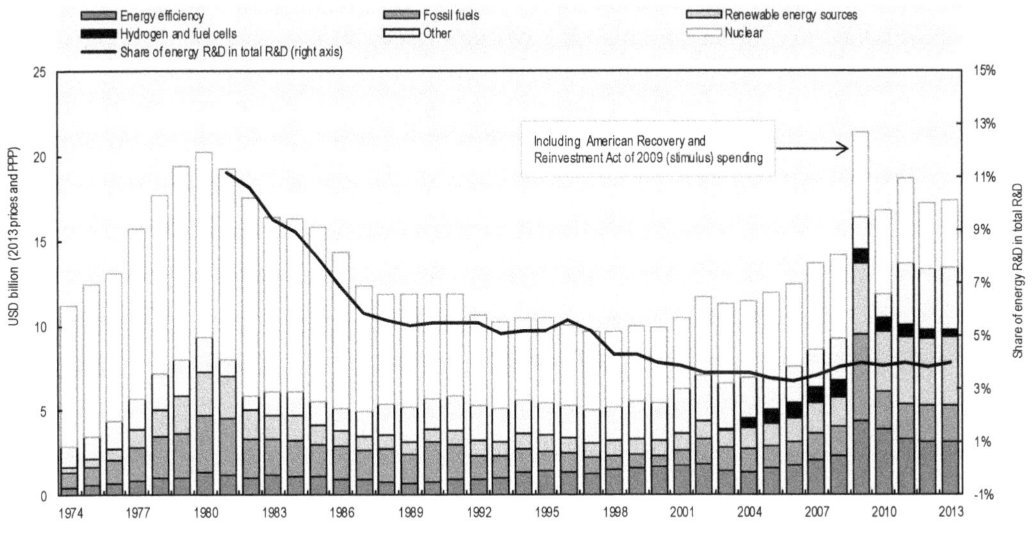

Source: IEA databases, 2014 cycle.

Private and public RD&D spending in activities outside the energy sector may also help the low-carbon transition, not to mention public spending outside IEA countries. This trend is therefore not the full picture of ongoing RD&D that may bring about breakthrough technologies for the low-carbon transition. It is nonetheless one useful indicator of these countries' intent to embark on the transition.

Moreover, R&D expenditures are an inadequate indicator of innovation, as they reflect inputs to innovation and not outcomes. However, evidence based on patent statistics – which suffer from their own shortcomings as a measure of innovation – indicates that there has been a global acceleration in technological progress in areas such as lighting, electric power, electric and hybrid vehicles, energy generation and electricity storage, observed in so-called patent bursts (OECD, 2014b). The penetration of these technologies and the emergence of new ones will require sustained innovation efforts. Policies in support of innovation deserve a closer look to ensure consistency with the low-carbon transition.

Inducing private R&D

Governments can choose among various tools to leverage business R&D. On the supply side, they can offer direct support via grants or procurement or use fiscal incentives such as R&D tax incentives. Indeed, more countries are now using tax incentives than in the past; in many countries the schemes are more generous than ever. R&D tax incentives are present in 28 OECD countries and also in Brazil, the People's Republic of China (hereafter "China") and South Africa.

Tax incentives for R&D are often considered to have certain advantages over direct support such as procurement or grants. As a market-based tool aimed at reducing the marginal cost of R&D activities, they allow firms to decide which R&D projects to fund. They do have shortcomings, however. Figure 4.2 shows a variety of country situations in terms of the generosity of R&D tax incentives. Depending on design, the benefits can be restricted. Most obviously, the benefits do not depend only on R&D expenditures, but also on profitability (e.g. in case of a credit on corporate income tax). This restrictive practice may be an impediment for small new companies that may need a few years to become profitable. However, some countries do allow firms to benefit from the subsidy when they are not profitable. For example, this can be addressed through the use of carry-over provisions, which allow loss-making firms to benefit from the credit once they become profitable (this is the case in Japan). It can also be addressed by allowing firms to claim an equivalent and immediate refund for unused credits or allowances, as in Australia, Canada, France and the United Kingdom.

Figure 4.2. **Tax subsidy rates on R&D expenditures, 2013**

1-B-Index, by firm size and profit scenario

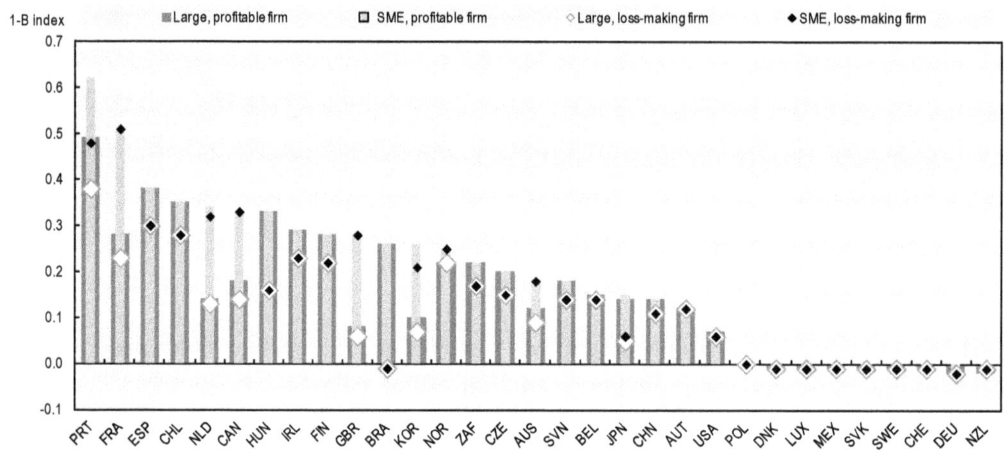

Note: SME: small and medium-sized enterprise. The B-index is a measure of the level of pre-tax profit that a representative company needs to generate to break even on a marginal, unitary outlay on R&D (Warda, 2001), taking into account provisions in the tax system that allow for special treatment of R&D expenditures. It is customary to present this indicator in the form of an implied subsidy rate, namely *one* minus the *B index*. More generous provisions imply a lower breakeven point and therefore a higher subsidy.

Sources: OECD R&D Tax Incentives Indicators; based on the 2013 OECD-NESTI data collection on tax incentives support for R&D expenditures, www.oecd.org/sti/rd-tax-stats.htm; OECD (2013c), *OECD Science, Technology and Industry Scoreboard 2013: Innovation for Growth*, OECD Publishing, Paris, http://dx.doi.org/10.1787/sti_scoreboard-2013-en.

To the extent that the low-carbon transition should encourage the birth and growth of many new entrants, there may be a policy misalignment to be corrected (i.e. in countries to the bottom right-hand of Figure 4.2). Solutions exist to support R&D in young loss-making firms via an immediate refund for expenditure on R&D personnel wages, or strengthening support for the commercialisation of public research and start-ups, and enhancing access to finance. Keeping in mind the importance of enhancing competition with innovative new entrant companies, the design of R&D tax incentives ought to be reconsidered to avoid a bias towards incumbents.

Irrespective, direct R&D support is an important complement to R&D tax incentives, allowing governments to target particular technological domains or firm types. The potential benefits of "targeting" raises its own problems, not least the identification of promising technologies and sources of innovation. Many countries focus on specific groups of firms, and particularly small and medium-sized enterprises (SMEs). In the OECD area, they carry out about one-third of total business R&D. In most countries, over 50% of direct R&D public funding (e.g. grants, loans or procurement contracts) goes to finance R&D for

SMEs. However, in Japan, Sweden, the United Kingdom and the United States about 90% of government direct funding goes to larger firms (see Figure 4.3).

Figure 4.3. **Business expenditures on R&D by SMEs and direct government funding, 2012**

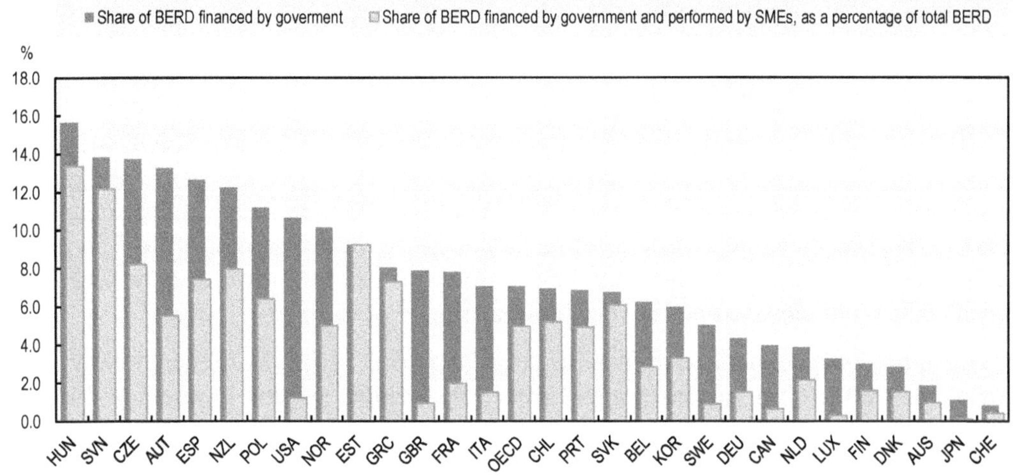

Source: OECD (2013d), "Business enterprise R-D expenditure by size class and by source of funds", *Research and Development Database*,**Error! Hyperlink reference not valid.** http://stats.oecd.org/Index.aspx?DataSetCode=BERD_SIZE (accessed March 2015).

In addition to the challenges associated with the identification of promising technologies, attention must be paid to the realisation of both short- and long-term policy objectives. On the supply side, support for basic research will remain important for radical (and riskier) innovations with potential long-term benefits. More generally, programmes that foster co-operation between firms (both small and large) with universities are likely to be important in bridging the gap between basic and applied research.

Demand-side innovation policy measures such as public procurement, information dissemination, advanced market commitments and technology prizes can be an important complement to supply-side measures (OECD, 2011a). The concept of fostering innovation through demand-side policy – particularly public procurement – is not new. Indeed, some countries have pursued active technology procurement policies for decades. In terms of climate objectives, this has been particularly important in the areas of energy and transport. Nonetheless, even though public procurement represents around 16% of gross domestic product (GDP) in OECD countries, little of this spending explicitly addresses innovation. More generally, there may be "policy failures" in public expenditure management which can prejudice climate-friendly innovation. Issues such as single-year budget frameworks, split responsibilities for capital and operating costs, and the nature of accounting procedures (i.e. treatment of capital charges) can result in investment in technologies which are more climate-intensive (OECD, 2003).

Facilitating entry and exit and the development of knowledge-based capital

> It will be essential to remove barriers to trade in clean technologies as well as to the entry of new firms, and to improve conditions for entrepreneurship, especially in light of growing evidence that young firms represent an important source of more radical innovations. (OECD, 2014b)

At the level of firms, innovation rests on a competitive environment. New firms are important vehicles through which much innovation is brought to the market, and firm entry and post-entry growth is essential. In the specific context of this report, this implies that carbon-intensive incumbents are also challenged to improve and change their technologies by newcomers providing low-carbon alternatives. This is best fostered by a policy environment that encourages risk-taking – i.e. does not unduly penalise failure. Against this backdrop, the sharply declining rate of start-ups in OECD countries over the past decade is a cause for concern (Criscuolo et al., 2014).

Differences in firm dynamics and demographics across OECD countries indicate where progress is possible. In particular, there are striking cross-country differences on the growth performance of young firms (Criscuolo et al., 2014). This is a function of incentives that encourage or discourage the allocation of resources to the most dynamic firms. The policy context is crucial. Figure 4.4 illustrates how estimated resource flows to patenting firms vary with different policy settings in OECD countries.

The results indicate significant benefits from increased access to seed and early-stage financing, as well as from increased efficiency in the judicial system. In addition, benefits arise from reforming product market regulations (PMRs) which inhibit competition, as well as reducing policy-induced barriers to exit (e.g. excessively strict bankruptcy codes). Reducing such barriers will accentuate competitive pressures, encouraging inefficient firms to exit and channel resources to firms that are best able to make use of the resources. The latter effect can also be encouraged by less stringent employment protection legislation. While these and other measures listed in Figure 4.4 are typically implemented with other policy objectives in mind, their unintended implications for innovation should be taken into account. In any event, recommendations in this area should also consider countries' broader social and economic context. For instance, less stringent employment protection legislation may be more politically acceptable in jurisdictions with higher unemployment benefits. There is also a broader question about an economy's ability to drive a transformative agenda such as the low-carbon transition without a strong social contract (see Galgoczi, forthcoming, for a discussion of the industrial transition of Germany's Ruhr region).

The above examples suggest that framework policies could significantly affect the extent to which patenting firms attract the tangible resources required to implement and commercialise new ideas. This can have important implications for climate innovation. For example, addressing failures associated with early-stage financing may have particular benefits due to the preponderance of cases in the climate domain which involve the co-existence of novel technologies with uncertain performance characteristics and young firms with little track record or collateral. Moreover, market and technology risk is compounded by policy risk, with important implications for access to finance. Considerable work has been undertaken on the central role of policy settings on the provision of finance for "clean" technologies in OECD economies.[3] However, more work is required to assess the effectiveness of targeted financial policy interventions (e.g. public investment in "green" venture capital funds) relative to more general policies associated with the functioning of capital markets.[4]

While a successful low-carbon transition should be accompanied by the birth – and growth – of new firms, the role of incumbents should not be strictly juxtaposed with those of entrants, as the former also produce radical innovation from their cumulative incremental innovations (as has been the case with LED technologies). In addition, some large incumbent companies such as Toyota have opened their patents in low-carbon technologies – in this case, fuel cells – to other firms to enable them innovate in the same area. Tesla announced in June 2014 that it would offer its proprietary technology to other car manufacturing companies with an interest in electric vehicles (Wall Street Journal, 2014).

Figure 4.4. **Framework policies and resource flows to patenting firms, 2003-10**

A. Change in firm employment associated with a 10% change in the patent stock

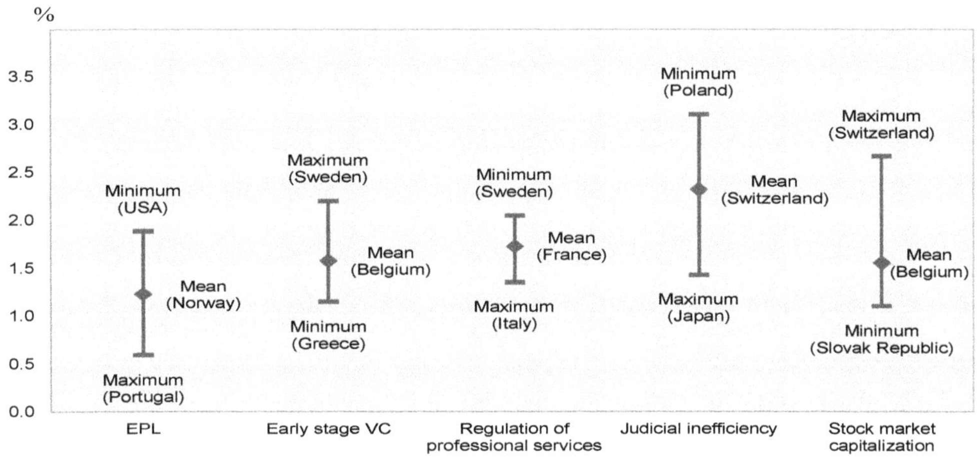

B. Change in firm capital associated with a 10% change in the patent stock

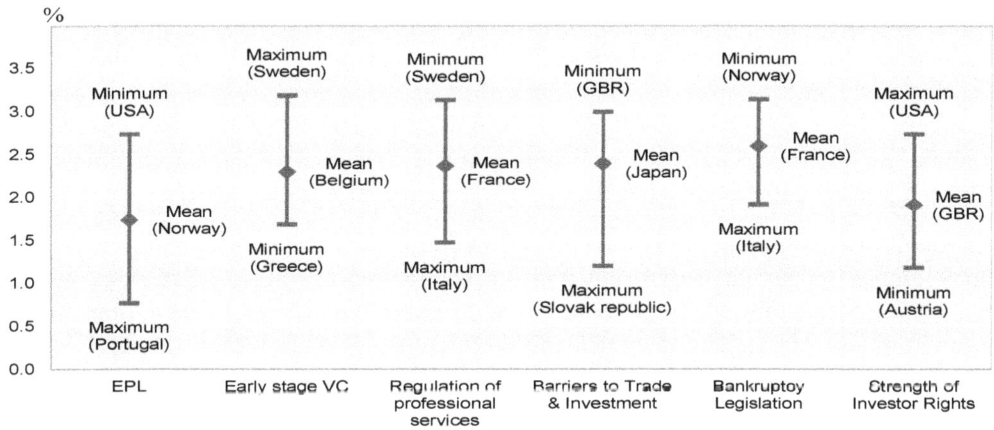

Notes: This figure shows that the sensitivity of firm employment and capital to changes in the patent stock varies according to the policy and institutional environment. To calculate policy effects, coefficient estimates are combined with the average values of the policy indicators for each country over the sample period. The labels "minimum" ("maximum") denote the country with the lowest (highest) average value for the given policy indicator over the sample period. EPL: Employment Protection Legislation; VC: Venture Capital.

Source: Andrews, D. and C. Criscuolo (2014), "Knowledge-based capital, innovation and resource allocation", *OECD Economics Department Working Papers*, No. 1046, OECD, Paris, available at: www.oecd.org/officialdocuments/publicdisplaydocumentpdf/?cote=ECO/WKP%282013%2938&docLanguage=En.

Environmental policy design can also influence firm dynamics and competition (Albrizio et al., 2014). Many environmental regulations are differentiated by plant vintage, with less stringent regulations for existing facilities relative to new facilities. Such "grandfathering" regulations are well documented in the case of power plants.[5] This generates a bias against investment in new sources ("new source bias"), but also discourages plant exit, as the opportunity cost of closure is too high.

Any regulatory scheme (e.g. health and safety rules, building regulations, access to networks, etc.) which favours incumbent firms or facilities relative to entrants will discourage exit, and to the extent that older vintage correlates with technologies with lower environmental performance there will be negative implications in terms of carbon emissions. In some countries the debate has gone further, with discussions about "payments for closure".[6] However, such a strategy raises the potential for strategic behaviour, with managers delaying closure in anticipation of receiving compensation for doing so.

Should government support specific technologies for the low-carbon transition?

The sheer breadth of domains which can generate environmentally benign innovations complicates the task of policy makers. The kinds of breakthrough innovations that can generate significant environmental benefits can come from fields as diverse as ICT, materials sciences and biotechnology. Breakthroughs could come from new infrastructures but also from innovations in the way existing systems are used (i.e. reconfiguration processes such as those underway with the use of smart meters in electricity systems). As noted above, this uncertainty strengthens the case for greater public investment in basic research, particularly with respect to transformative technologies.

However, in practice in a number of fields, governments have made policy bets on specific technologies in a more directed manner, including on the basis of their low-carbon attributes, assumed future contribution to reducing emissions, and their potential for future cost reductions – analyses such as the IEA Technology Roadmaps facilitate the positioning of today's technology policy efforts in a longer run process towards large-scale deployment.[7] This has been most visible in the power generation sector, with government programmes in support of nuclear, subsidies to wind farms and solar photovoltaic (PV) installations, carbon capture and storage, or infrastructure for electric vehicles.

Although increasingly ambitious technology-neutral market signals should be the core of future climate policy developments if countries are serious about addressing climate change, governments are likely to continue supporting the research, development and deployment of specific technologies that can facilitate the transition to a low-GHG world. This can be justified by factors such as imperfect commitment and path dependency, for instance in the case of infrastructure supporting the deployment of breakthrough technologies. This is illustrated by the findings of Acemoglu et al. (2012) on the economic efficiency advantage of directing technical change towards low-carbon paths in response to climate change.

It is therefore legitimate to ask whether it is possible to identify technologies that are most likely to deliver the needed breakthroughs, and as such represent a good investment of public monies. Given the hazards associated with identifying promising technologies at an early stage of development, governments often support a "portfolio" of technologies, hedging their policy bets in the face of uncertainty.

This still begs the question of which technologies to include in the portfolio. One possible risk-averse strategy is to support innovation which has broad potential applications. For example, OECD work has indicated that support for research in energy storage and grid quality was very valuable in terms of patents generated, compared with support for specific renewable energy technologies. Arguably, there has been a temporal misalignment with insufficient attention paid to the technological characteristics of the supporting technologies required to accommodate the rapid increase in the penetration of intermittent renewable energy (Johnstone and Haščič, 2012).

More generally, a methodology has been developed to "uncover attributes of inventions (as reflected in patent data) which may serve as "leading indicators of subsequent technological and market development" (OECD, 2014b). Such indicators include the breadth of industrial sectors in which the technology is subsequently applied (industrial generality), the extent to which it builds on previous inventions (radicalness), the spatial breadth of the markets in which it is protected (patent family size) and the degree to which it draws upon basic rather than applied research (closeness to science). If robust, this can help policy makers to prioritise public support in various technological fields.

However, provision of targeted support for specific technologies requires continuous evaluation, not least as any direct policy support raises the question of rent-seeking behaviour by beneficiaries. This can eventually lead to excessive government expenditures whenever support is granted that is no longer justified because the policy objective has been met or, to the contrary, when support has not delivered the expected technological breakthrough. Some effort is needed in innovation policy to avoid such situations, e.g. with the development of criteria to determine whether beneficiaries are in one or the other situation. Sunset clauses can also be useful, but can still be open to negotiation and lobbying by rent-seeking companies. It is preferable to develop clear and non-discretionary criteria for policy exit, and to communicate these criteria *ex ante* so that beneficiaries can foresee future policy developments.

The governance of innovation

An important alignment challenge in the area of climate innovation arises from the fact that the benefits associated with the development and adoption of technologies cross borders. The externality is trans-frontier, and as such there is an additional channel through which incentives may be misaligned which is absent from other domains. Moreover, the areas of the world in which the potential benefits of their adoption are greatest may not be the sources of innovation. This has been shown to be the case in terms of innovation related to climate adaptation (Dechezleprêtre et al., 2015).

This raises issues of international co-ordination if global challenges – including climate change – are to be addressed (see Chapter 5 of the *OECD Innovation Strategy*, forthcoming). OECD analysis of the challenges in establishing international research infrastructures, including shared financing, governance and legal frameworks, suggest there is no one-size-fits-all model for such facilities. Increasingly, and partly a consequence of the broader move towards more open science and innovation, international distributed research infrastructures have become more prevalent. These are located across several countries, share a common purpose and are co-ordinated in some way, but otherwise can be of very variable geometry.[8] The inclusion of countries with weaker science, technology and innovation capacities as full partners is necessary and may require specific actions to build capacities (Poirier et al., 2015).

Issues of co-ordination and alignment also arise within individual countries. This is hardly surprising, as both fostering innovation and addressing climate change are issues requiring policy responses that do not fit neatly with the competencies of any single governmental department or agency, or even with a single level of government. Multi-level governance requires that authorities at various levels possess not only the right capabilities and resources to effectively run their own competencies, but also the capacity and means to enter into negotiations, align their policies and conclude agreements with authorities from other levels.

Another complementary dimension is that of co-ordination between policy domains at the same level of government, i.e. horizontal governance. One specific example of the need for better governance related to innovation with climate implications concerns systems innovation. Although many national governments have included climate objectives in their economic development strategies, achieving this goal will require wide-ranging changes to transport, energy, buildings and other systems. The "smart city" initiatives that mobilise technological and social innovations to make the production and consumption of a city's goods and services less climate-intensive illustrate this point. At the national level also, improved governance mechanisms and better means of engaging a range of stakeholders are needed to facilitate system innovation. Finland and the Netherlands, for example, have developed public-private partnerships to foster co-ordination and alignment.[9]

A final issue related to governance which can affect innovation in the climate domain is the treatment of risks above and beyond those associated with climate risks themselves. Uncertainty in science policy is pervasive. Innovation in some of the areas which have the potential to yield significant potential climate benefits may also carry important potential risks (i.e. biotechnology, nanotechnology). R&D is a probabilistic investment and the direction of science and innovation and their ramifications involve elements of randomness and often long-term and hard-to-foresee outcomes. Governments need to deploy a range of tools – foresight studies, technology assessments, formal cost-benefit analyses – to help steer policy and reduce unnecessary risks. But all such techniques have strengths and weaknesses. Getting the balance right is an open research question, and ongoing discussions about providing support for geo-engineering technologies is a reflection of the need for further work in this area.

Putting the labour markets at work for the low-carbon transition

> To the extent that it can be predicted, it does not appear that the transition towards green growth is likely to imply rates of labour reallocation or rates of change in job skill demands that are outside of historical experience. However, that conclusion may say more about how difficult it is to predict the labour market consequences of decoupling economic growth from harmful environmental impacts than how easily green growth-driven structural change can be managed. (OECD, 2012b)

Can we anticipate skills needs for the low-carbon transition?

The transition to a low-carbon economy requires a parallel evolution of skills and shifts in the labour force. Under a 2°C Scenario, most carbon-intensive activities will have decarbonised or will have been out-competed by low-carbon substitutes. The following are illustrations of changes with implications on skills and labour:

- The extraction of coal, oil and gas should be reduced significantly in volume from today's levels in the course of the century, even if a successful carbon capture and storage technology could make their use more climate-friendly. These sectors have already witnessed impressive labour productivity gains in the past decade, but the low-carbon transition could result in further labour force reductions in decades to come. Alternative energy sources (e.g. renewables) should witness very rapid employment growth rates, albeit from a relatively small base.

- The imperative of more efficient buildings implies the use of new materials and skills. In most OECD countries, the deep renovation of existing buildings could represent an important new market for the construction sector (OECD, 2012b).
- In the automotive sector, the possible penetration of the electric engine would trigger change of skillset in downstream activities (car repairs and mechanics). Lighter vehicles to improve fuel economy will require the use of new materials (lighter ultra-strength steel, aluminium, plastics or carbon fibre), with new skills needed on the factory floor and in repair shops.
- The circular economy, often presented as an important structural change to improve resource efficiency and lower greenhouse gas emissions, may also create new jobs, but eliminate jobs in the extraction of virgin resources.
- Last but not least, skills policies also need to support the innovation process itself (e.g. workers with advanced science, technology, engineering and mathematics skills (STEM).

The labour force and skills will have to accompany these and similar changes in a range of activities. One of the preconditions for a smooth decarbonisation of the overall economy is that the vocational education and training system adapt in a timely way to changing skill demands. Another is that labour market institutions and policies support the necessary reallocation of labour from shrinking to growing firms and activities. This is all the more important as there are concerns about job destructions related to the shift to a low-carbon economy connected to threats of carbon leakage (see Chapter 1).

Is there a policy alignment issue?

The fundamental alignment question is whether labour markets and education and training systems will be able to respond to the new demands of the low-carbon economy, or whether governments should intervene to remove barriers to change. The analysis is complicated by the uncertainty on future evolutions: which new activities will want to enter the market, and what will be the demand on the labour force to develop them?

Good practice in employment policy could also support the low-carbon transition:

> [...] countries which have better reconciled labour mobility with income security (achieving so-called flexicurity) are likely to have an advantage managing an efficient transition, while also limiting political opposition to green growth policies grounded in concerns about the economic dislocation they could imply. Similarly, countries with education and training systems that perform particularly well in developing high-level STEM skills are likely to be better placed to become leaders in eco-innovation, while national systems of continuing vocational training that are particularly responsive to employers' evolving job skill requirements may help to foster faster and more efficient diffusion of green technologies and working practices across the economy. However, there is as yet little systematic evidence on these questions. [OECD, 2012b]

The evidence on job shifting from high- to low-carbon activities is scant at present and probably not representative of economy-wide changes. For instance, "spotty evidence suggests that a significant share of the conversion of the electricity sector from fossil fuels to renewable sources is occurring within large electrical utilities, a number of which are actively retraining their workforces as part of their implementation of a transition to clean energy" (ibid.). In this example, companies adjust to the new market opportunities resulting from the deliberate action by governments to support low-carbon technologies; no particular public policy change seems required to accompany this change in workforce competence. One can easily think of industrial basins and communities with a fossil fuel basis, where lower demand would have visible disruptions and could need accompanying measures.

There are instances, however, of policy-driven new markets that stumble over a lack of appropriately skilled workers or the absence of a skills certification scheme that allows employers and customers to distinguish between workers who have the required capabilities and those who do not (Box 4.1). Training programmes can be put in place to overcome such problems when identified. These issues are more likely to occur in activities characterised by a large number of SMEs with less training capacity, such as the construction sector.

Box 4.1. If you build it, they will come? Certification in the construction sector

Policy makers are increasingly mindful that skills gaps may undermine the effectiveness of certain policies: in and of itself, the creation of a new market by policy does not guarantee that an adequately skilled workforce will emerge to serve this market in a timely manner. Such policies should be devised with a parallel assessment of required skills sets and their availability. Certification can be an instrument to ensure that the workforce on the supply side of a policy-driven market is adequate.

The skills of construction workers is often pointed out as a possible weak link in the implementation of building codes introduced to improve the energy efficiency of housing and commercial buildings. France introduced a certification system with a specific label for companies that conducted specific training of their staff on energy efficiency (the RGE label, which stands for recognised environmental guarantee). Several public awareness campaigns run by the French energy efficiency agency (ADEME) encouraged consumers to ask for trained professionals with RGE labels. Since 2014, public financial assistance for renovating buildings is conditioned to the employment of RGE-labelled companies. In 2008, France also launched FEEBAT, a programme of continuous vocational training partly financed with a contribution of Électricité de France, a French utility, through the white certificates system. Between 2008 and 2012, the scheme trained 48 000 construction professionals. The programme has been prolonged until 2017.

The United Kingdom has established licensing programmes to address a market barrier to the renovation of buildings, specifically in response to its Green Deal programme. Through the UK Accreditation Service, criteria are developed for the accreditation of Green Deal assessors which builds on pre-existing qualifications, and of Green Deal installers. The profession sees the Green Deal certification "as a way of driving 'cowboys' out of the industry" (Lane and Power, 2010, in OECD/Cedefop, 2014).

Source: OECD/Cedefop (2014), *Greener Skills and Jobs*, OECD Green Growth Studies, OECD Publishing, Paris, http://dx.doi.org/10.1787/9789264208704-en.

Given the many uncertainties on the sectoral and regional evolutions that the low-carbon transition will generate, the possible challenge for employment policy is to be able to identify and quantify both the creation of jobs requiring new skills,[10] and to provide appropriate incentives for training where it may be lacking. While this can be seen as a good policy across the board, the fact that governments, not markets, are giving the primary impetus to the low-carbon transition puts responsibility on governments rather than market forces alone, to ensure a proper labour market response.

There is policy experience in this area that could be replicated in other countries to anticipate workforce and skills challenges in their transition to low carbon (Box 4.2). The European Commission established an initiative to train Europe's workforce for the energy efficiency renovation of buildings (EC, 2014). BUILD UP Skills started from the observation that craftsmen and on-site workers are not always trained to properly implement energy efficiency and renewable energy solutions in the building sector. This initiative started with country-level evaluations of skills needs and gaps, revealing that more than 3 million workers in Europe will require training by 2020, most of it in the form of continuous vocational education for existing workers. In a second phase, BUILD UP Skills will extend funding for large-scale qualification and training schemes. The skills gap identified in this programme is a signal to other countries that low-carbon improvement measures in the buildings sector must be accompanied by information and training programmes in the construction sector.[11]

Policy misalignments restricting innovation in industry

Innovation may arise from, or be hampered by, detailed regulations applying to economic activities. Regulations may have been established some time ago on the basis of past technical knowledge that may now be outdated, and will not automatically catch up with policy priorities such as climate change or technological progress. In other instances, regulations favour incumbent technical solutions and implicitly bar competition. While core climate policy instruments should spur innovation, certain regulations clearly stymie it. This section examines some policy misalignments (in particular regulatory barriers) associated with the flow of resources in, out and within the industrial sector.

Improved resource efficiency – for the use of energy, materials, water, biomass, waste and land – is fundamental to reducing industrial sector GHG emissions. Market forces and many government policies already encourage resource efficiency in individual industrial facilities, but there are often missing incentives and sometimes disincentives – arising from policy misalignments – that inhibit wider-scale efficiencies involving interactions among plants in different activities. Regulations governing industrial and end-use products, by-products and wastes (of all resource types: energy, materials, etc.) can sometimes preclude their most efficient use in the industrial system.

The potential GHG reductions from efficiency improvements are particularly pronounced when their resource flows are associated with the production of basic materials such as cement, steel, chemicals and paper. The products, by-products and wastes (and the resource streams they compete with) of these industries all contain large amounts of embedded energy and GHG emissions, so resource efficiency gains in these activities can result in proportionally large GHG emission reductions.

This section begins with two particular examples from the cement and concrete sectors. Not because cement and concrete production are particularly resource inefficient, but because their production processes have great flexibility in the use of raw materials (including by-products from other industries) and energy (including many alternative fuels). IEA (2012) provides estimates of the GHG emissions reductions of two of the items considered in this chapter – alternative fuels and clinker substitutes in the cement sector (Figure 4.5).

Box 4.2. Skills anticipation

Is the workforce properly trained to implement the various changes that the low-carbon transition requires? If not, can governments play a constructive role in anticipating emerging skills needs and meeting them in a timely fashion? The OECD has launched an inquiry in labour ministries to find out how they identify skills needs and take measures to remedy possible gaps, including whether these initiatives take account how environmental policy will change future job skill requirements. While all OECD countries attempt to be forward-looking in their educational and training policies, only six report having undertaken substantial efforts to forecast and meet new job skills requirements related to the low-carbon transition: Austria, Canada, France, Greece, Ireland and Italy. To date, three types of approaches have been followed, all of which have the potential to minimise the misalignment between skills and the green transition.

- **Episodic studies of green economy skills.** Canada and Ireland have conducted occasional studies of the emerging skills requirements associated with the transition towards a low-carbon economy. Ireland carried out a 2010 study of the skills requirements of firms working in the "green economy sector" (Expert Group on Future Skills Needs, 2010), and information from this study fed into Ireland's Action Plan for Jobs (Government of Ireland, 2013). Canada's low-carbon goods and services activities indicated that they have experienced or anticipate a lack of skilled labour (NRTEE, 2012).
- **Co-ordination of vocational education and training (VET) planning with the implementation of environmental policies.** In Italy, local governments regularly adapt VET programmes at the post-secondary education level in light of new job skill requirements stemming from environmental policies (e.g. to enforce regulations on environmental quality and intervene in the management of waste water).
- **Systematic forecasting of emerging demands for "green skills".** The Austrian and French governments regularly forecast the future skill requirements resulting from the transition to low-carbon growth and feed this information into education and training policy:
  - The Austrian New Skills framework monitors trends in skill demands via its Standing Committee on New Skills. This group recently concluded that "energy efficiency, the use of alternative sources of energy, resource-saving production, the identification of cause-effect relationships, recycling and ecologically sound disposal of materials, etc. are increasingly becoming a business necessity for all companies. In this connection the development of environmentally efficient production processes, service processes and products constitute an increasing challenge for employees from many different areas and at all qualification levels, both technically and in terms of awareness-raising and understanding" (Bliem et al., 2011: 2)
  - In line with the national strategy and national pact to transition towards a green economy, France has established *L'Observatoire national des emplois et métiers de l'économie verte* (National Observatory for Green Economy Jobs and Skills), which forecasts the sectoral and macroeconomic impact of the green transition, with special attention to its implications on the numbers of jobs and skills requirements (Commissariat Général au Développement Durable, 2014). In developing these forecasts and assessing their implications for training policies, partnerships have been developed with trade unions, employer organisations and Pôle Emploi (the public employment service). Pôle Emploi, for example, has studied the supply and demand for green skills to guide the design of its programmes to up- or re-skill job-seekers to better meet the requirements of this transition (Pôle Emploi, 2011a; 2011b).

As part of the initiatives of the G8 Deauville partnership, Tunisia is implementing a pilot to address green skills needs in the context of a youth employment programme. Special attention is being paid to the creation of clean tech solutions in areas such as renewable energy, waste and recycling, agriculture, tourism and construction, and training is being organised in anticipation of "green jobs" in these sectors.

Sources: OECD (2015b), *Anticipating and Responding to Changing Skill Needs*, OECD Publishing, Paris; OECD (2015c), *Investing in Youth Tunisia: Strengthening the Employability of Youth During the Transition to a Green Economy*, OECD Publishing, Paris, http://dx.doi.org/10.1787/9789264226470-en.

Possible misalignments are also identified in two more general industry examples outside of the cement sector: the development of industrial symbiosis and increased deployment of combined heat and power (CHP).

Figure 4.5. **Technologies for reducing cement direct CO_2 emissions between the 4DS and 2DS**

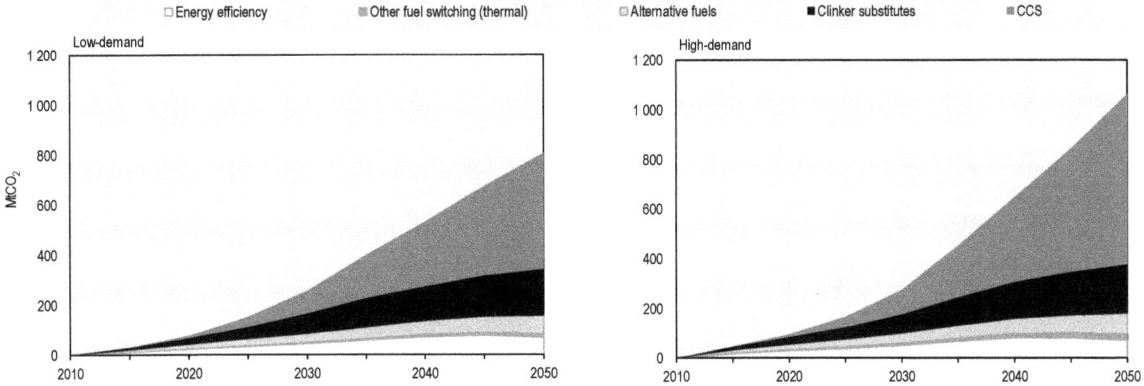

Notes: CCS: carbon capture and storage. Alternative fuels account for 22-24% of total cement sector emission reductions between 2DS and 4DS scenarios (4DS and 2DS refer to the IEA 2°C and 4°C scenarios for GHG mitigation) in 2020, 15% in 2030 and 8-9% in 2050. Clinker substitutes account for 33-34% in 2020, 29-32% in 2030 and 19-23% in 2050. Together, these measures could reduce CO_2 emissions by 45-46 $MtCO_2$-eq in 2020, 117-118 $MtCO_2$-eq in 2030 and 260-282 $MtCO_2$-eq in 2050.

Source: IEA (2012), *Energy Technology Perspectives 2012: Pathways to a Clean Energy System*, OECD/IEA, Paris, www.iea.org/publications/freepublications/publication/industry2009.pdf.

Case study: Co-processing energy needs for cement production

The majority of cement kilns are fuelled by coal, but other fossil and alternative fuels can also be burned (IEA/WBCSD, 2009). While alternative fuels have a lower CO_2 intensity depending on their exact composition (Sathaye et al., 2011) and can lead to reduced overall CO_2 emissions from the cement industry (CEMBUREAU, 2009), their use can also increase overall energy use per tonne of clinker produced if the fuels require pre-treatment such as drying (Hand, 2007).

Where alternative fuels are used at high substitution rates, tailored pre-treatment and surveillance systems are needed. Municipal solid waste (MSW), for example, needs to be screened and processed to homogenise calorific values and feed characteristics.[12] A well-designed regulatory framework for waste management is an important factor in facilitating the use of waste (IEA, 2009).

Waste fuels have been used in cement production for the past 20 years in Europe, Canada, Japan and the United States (GTZ/Holcim, 2006; Genon and Brizio, 2008); the Netherlands and Switzerland use 83% and 48% waste, respectively, as a cement fuel (WBCSD, 2005). Outside the OECD, the use of alternative fuels is not widespread. In developing countries, although interest is growing, substitution rarely exceeds 1% of the cement industry's fuel needs. There is significant scope to increase waste substitution globally with benefits for profits and the environment (IEA, 2009). Many developing countries such as China and nations in Southeast Asia are initiating programmes to promote co-processing of wastes in the cement industry. However, regulations, standards and the technical infrastructure in these developing countries are less mature than in those that have a long experience with co-processing waste in the cement industry (Hasanbeigi et al., 2012).

It is important that wastes are burned in accordance with strict environmental guidelines to avoid adverse environmental impacts such as extremely high concentrations of particulates in ambient air, ground-level ozone, acid rain and water quality deterioration (Karstensen, 2007; Schorcht et al., 2013). If MSW and sewage sludge are co-processed correctly and according to stringent environmental and emissions standards and regulations, there are no additional health and environmental risks compared to those that result when coal is used as a fuel (Rovira et al., 2011; Zabaniotou and Theofilou, 2008; Karstensen, 2007, 2008).

Hasanbeigi et al. (2012) examined international best practice in this area and found that effective regulatory and institutional frameworks are critical to ensure that cement industry co-processing practices do not have negative health or environmental impacts (see also Schorcht et al., 2013). The research also enumerated barriers to co-processing of MSW and sewage sludge in the cement industry, some of which could be streamlined:

- Permitting; Although the cement industry prefers uniform emissions standards for co-processors rather than case-by-case permitting of waste co-processing at plants, case-by-case permitting is necessary for hazardous waste, for environmental and health reasons.
- Regulations and standards: Some countries lack specific regulations and standards for co-processing of waste in the cement industry. Inadequate enforcement of waste management regulations in many developing countries is also one of the key barriers.
- Supportive policies: In many cases, co-processing may not be financially viable if its larger societal (waste management) benefits are not taken into account. Municipalities and governments that wish to pursue co-processing should design programmes and incentives based on co-processing's full benefits to the local community and environment.
- Public acceptance: Local residents and groups often perceive waste co-processing to be the same as waste incineration and automatically resist co-processing of MSW and sewage sludge in cement kilns. The major concern is usually the emissions from waste combustion, especially dioxin. Authorities should openly and publicly communicate emissions monitoring data and information from co-processing cement plants.
- Infrastructure: For example, existing infrastructure for sewage sludge is largely based on applying sludge to land or landfilling. Alternative infrastructure is needed for transport and pre-processing to cement plants.
- Lack of qualified workforce: The co-processing of waste in cement plants requires highly qualified experts to install and set up the equipment and trained personnel to operate the equipment. This capacity is presently limited in most developing countries.

Clinker substitutes in cement and concrete production

Clinker production is the most energy- and GHG emissions-intensive stage of cement making. Yet clinker can be substituted with materials that are by-products of coal power plants (fly ash) or iron and steel facilities that would lower GHG emissions from cement manufacturing. There are wide differences in the production and use of clinker substitutes (i.e. blended cements) around the world (Table 4.1). South America, India, China and Europe are the largest users of clinker substitutes. Cement makers in the United States use relatively little clinker substitution (at the cement level), but do most of their blending in concrete, not cement. This variation is due to historical preferences among the architectural and construction professions, which are slow to change. They also result from differing regulations on material specifications.

Table 4.1. **Clinker-to-cement ratios (inverse measure of clinker substitute use) by region**

In percent

	1990	2012	Share of cement production included in statistics, 2012
Africa	89.4	76.6	44
Asia (excluding China, India, CIS) + Oceania	89.6	80.2	36
Brazil	80.1	66.5	78
Central America	84.3	73.2	77
China (People's Republic of)	79.6	72.0	4
Commonwealth of Independent States (CIS)	81.5	79.6	21
Europe	78.6	72.7	94
India	86.2	70.5	47
Middle East	84.9	80.6	12
North America	89.9	82.1	78
South America excluding Brazil	76.5	69.5	54

Source: Cement Sustainability Initiative (2015), *Getting the Numbers Right database*, www.wbcsdcement.org/GNR-2012/geo/GNR-Indicator_3213-geo.html (accessed 8 January 2015).

The US Environmental Protection Agency (EPA) explored the benefits and barriers to increased use of clinker substitutes in federal construction projects. Its report found that use of recovered mineral substitutes (RMC) yields positive environmental benefits through lower resource consumption. It also found that the main regulatory barriers to increased uptake were prescriptive technical standards for cement, air pollution regulations and solid waste regulations (US EPA, 2008).

In the United States, there is insufficient recognition, by states' Departments of Transportation, of performance-based (as opposed to prescriptive- or ingredients-based) technical standards for cement. Prescriptive standards favour status quo materials, at the expense of innovative blends that would also bring about lower CO_2 emissions. In some cases, regulations governing an activity that produces a substitute create barriers as well (i.e. mercury and NO_x control in power generation in the United States, which affects the ability to use fly ash from coal plants; the US EPA is looking at technology options to overcome this barrier).

There are no uniform, national regulations in the United Sates for the beneficial use of recovered materials such as fly ash and blast furnace slag. Each state has its own solid waste regulations. Although many states act to facilitate the use of these materials in concrete, some state solid waste regulations governing the management of recovered materials may make it more difficult to beneficially use these materials.

Industrial symbiosis

Industrial symbiosis involves sharing the use of resources and by-products amongst industrial actors on a commercial basis through inter-firm recycling linkages. In industrial symbiosis, traditionally separate industries engage in an exchange of materials and energy through shared facilities (Beltramello et al., 2013). Successful co-siting and reuse of by-products as inputs provides clear environmental benefits through improved resource efficiency, leading to lower overall GHG emissions. Economic and social benefits can also be identified.

A well-known example of industrial symbiosis is Kalundborg, Denmark, where the first exchanges took place in the 1970s. Two other key examples are described in Beltramello et al. (2013): Kwinana Regional Synergies Project in Australia and Ulsan in Korea. In both cases, demand seems to have played an important role in kick-starting the symbiotic relationship. Demand came from both direct users of the industrial symbiosis process, i.e. companies wishing to save costs and generate new revenue streams, and from local communities concerned about local air and water emissions.

Policy has also played an important role in establishing the site. Beltramello et al. (2013) note:

> In the case of the Kwinana industrial symbiosis project in Australia, regulatory pressure from the Department of Environment and Conservation (DEC) and the Water and Rivers Commission was considered as a key factor. The air emission quality and water discharge quality standards helped to encourage the adoption of the innovation... However, regulations were also considered barriers that prevented or delayed the implementation of synergies, e.g. existing water and energy utility regulations and environmental regulations requiring intensive approval procedures for by-product reuse. (Beltramello et al., 2013)

Salmi et al. (2011) found similar barriers when investigating the potential of organising a system of metallurgical industries in an industrial symbiosis around the Gulf of Bothnia between Finland and Sweden. There are potentially three main forms of symbiosis to be found between the industries involved (carbon and stainless steel mills, a zinc plant and iron regeneration), none of which faces significant technological challenges.

However, developing this symbiosis would require changing the status of material from non-waste to by-product status. Typically this requires the company to apply for a new environmental permit, which can be a lengthy process. For example, a steelmaking plant in Finland needed six years to remove the waste status of slag and scrap metals. Article 5 of the EU Waste Framework Directive sets out four cumulative conditions for a substance or object resulting from a production process to be considered as a by-product, including that it can be used directly without further processing other than normal industrial practice.

Industrial combined heat and power

In many industrial situations, electricity, heat (steam), and sometimes cooling can be produced via a single process – combined heat and power (CHP), or cogeneration – much more energy efficiently and less emissions-intensively than via two (or three) separate processes. As such, many governments actively promote, and some even mandate, the implementation of CHP as a fundamental energy efficiency and climate change mitigation measure in many industries (China in the cement sector, for instance).

The economic attractiveness of CHP depends on an individual plant's own needs for electricity, heating and cooling; the external markets for any excess production of these energy streams; the quantity and quality of waste heat available; the prevailing tariffs for electricity and fuel (usually natural gas, coal or biomass); and the cost of capital. Some of these factors – especially access to external markets and energy tariffs – can be heavily influenced by policy. Examples where policies make the implementation of CHP difficult include (IEA, 2013, 2014; Kowalczyk, 2009):

- In India, lack of centralised data – on the amount of installed capacity, its distribution among sectors, the types of fuel used, the amount of power and heat/cooling generated – to help understand the potential for CHP and required efforts to support a substantial deployment programme. There is also a lack of pipeline networks for delivering heat/cooling to external users.
- In Korea, a hindrance to CHP deployment is misalignment in regulated natural gas prices (set high to recoup losses in the liquefied natural gas market) and electricity tariffs (set below the full cost of electricity generation).
- In the United States, emissions limits based on heat input or exhaust concentration can penalise CHP for its on-site generation of both electricity and thermal energy (as opposed to output-based emissions limits (emissions per unit of electricity or thermal energy output, which would support CHP).
- Lack of uniformity among states and regions in grid interconnection standards makes it difficult for equipment manufacturers to design and produce modular packages that can easily connect to the grid in various applications in the United States.
- In the United States, a variety of barriers relating to the treatment of CHP by utilities have been identified. These include utility tariff structures, excessive charges by local utilities for future lost revenue or as compensation for continued access to the distribution networks.
- Disparate tax depreciation policies for CHP installations which may discourage CHP project ownership arrangements, thereby increasing the difficulty of raising capital and discouraging development in the United States.

Revisiting industry regulations

In light of the narrow scope adopted in what precedes and of the identified misalignments, every opportunity should be taken to revisit regulations applying to industrial activities that represent large GHG sources and to screen these regulations for their inadvertent consequences on low-GHG innovations. This is, no doubt, a significant task for regulatory and standard-setting bodies, but is necessary if climate policy signals are to trigger the intended technical changes. In the alternative, a number of industrial activities will be hindered in their ability to contribute to climate change mitigation.

Notes

1. 1.1% a year against 4.2% earlier, although it accelerated in 2012.
2. See also OECD (2013b) on the contribution of biotechnology.
3. See Cárdenas Rodríguez et al. (2014), Criscuolo and Menon (2014), and Criscuolo et al. (2014). See also OECD (2015a).
4. See Lerner (2011) for a discussion.
5. For example see Ellerman (1998); Levinson (1999); Bushnell and Wolfram (2007).
6. See Riesz et al. (2013).
7. See: www.iea.org/roadmaps for technology specific roadmaps as well as guidelines for countries seeking to develop their own roadmaps.
8. The OECD has facilitated the establishment of several such distributed infrastructures, most recently in relation to Scientific Collections (SciColl, 2013), and analysed the lessons learnt, including different governance options (OECD, 2014c).
9. Strategic centres for science, technology and innovation (SHOKs) in Finland and the Top Sectors approach in the Netherlands.
10. For example, OECD (2012b) already points out a lack of green ICT-related skills in many OECD countries.
11. Another example of a response to a policy-driven market creation is the training organised by the regional government of Extremadura, Spain to provide renewable energy specialists and project managers as it was rapidly developing its solar energy capacity. The solar programme created approximately 3 000 jobs.
12. The high temperatures and sufficiently long residence time in cement kilns and other characteristics of cement production make co-processing of waste materials a viable strategy (Hasanbeigi et al., 2012).

References

Acemoglu, D. et al. (2012), "The environment and directed technical change", *American Economic Review*, American Economic Association, Vol. 102, No. 1, February, http://dx.doi.org/10.1257/aer.102.1.131.

Albrizio, S. et al. (2014), "Do environmental policies matter for productivity growth? Insights from new cross-country measures of environmental policies", *OECD Economics Department Working Papers*, No. 1176, OECD, Paris, available at: www.oecd.org/officialdocuments/publicdisplaydocumentpdf/?cote=ECO/WKP%282014%2972&docLanguage=En.

Andrews, D. and C. Criscuolo (2014), "Knowledge-based capital, innovation and resource allocation", *OECD Economics Department Working Papers*, No. 1046, OECD, Paris, available at: www.oecd.org/officialdocuments/publicdisplaydocumentpdf/?cote=ECO/WKP%282013%2938&docLanguage=En.

Beltramello, A. et al. (2013), "Why new business models matter for green growth", *OECD Green Growth Papers*, No. 2013/01, OECD Publishing, Paris, http://dx.doi.org/10.1787/5k97gk40v3ln-en.

Bliem, W. et al. (2011), "AMS Standing Committee on New Skills 2010/2011: Report on the results of the expert groups", *AMS Info*, Vol. 211.

Bushnell, J.B. and C.D. Wolfram (2007), "The economic effects of vintage differentiated regulations: The case of new source review", UC Berkeley, Center for the Study of Energy Markets.

Cárdenas Rodríguez, M. et al. (2014), "Inducing private finance for renewable energy projects: Evidence from micro-data", *OECD Environment Working Papers*, No. 67, OECD Publishing, Paris, http://dx.doi.org/10.1787/5jxvg0k6thr1-en.

Cedefop (2013), "Skills for a low carbon Europe: The role of VET in a sustainable energy scenario", Synthesis Report, *Research Paper No. 34*, European Centre for the Development of Vocational Training, Publications Office of the European Union, Luxembourg, available at: www.cedefop.europa.eu/en/publications-and-resources/publications/5534.

CEMBUREAU (2009), "Sustainable cement production: Co-processing of alternative fuels and raw materials in the European cement industry", CEMBUREAU, Brussels, available at: www.cembureau.eu/sites/default/files/Sustainable%20cement%20production%20Brochure.pdf, cited in IPCC (2014).

Cement Sustainability Initiative (2015), *Getting the Numbers Right database*, www.wbcsdcement.org/GNR-2012/geo/GNR-Indicator_3213-geo.html (accessed 8 January 2015).

Commissariat Général au Développement Durable (2014), *Observatoire National des Emplois et Métiers de l'Économie Verte: Rapport d'Activité 2013*, ministère de l'Écologie, du Développement durable et de l'Énergie, Paris.

Criscuolo, C. and C. Menon (2014), "Environmental policies and risk finance in the green sector: Cross-country evidence", *OECD Science, Technology and Industry Working Papers*, No. 2014/01, OECD Publishing, Paris, http://dx.doi.org/10.1787/5jz6wn918j37-en.

Criscuolo, C. et al. (2014), "The dynamics of employment growth: New evidence from 18 countries", *OECD Science, Technology and Industry Policy Papers*, No. 14, OECD Publishing, Paris, http://dx.doi.org/10.1787/5jz417hj6hg6-en.

Dechezleprêtre, A. et al. (2015), "Invention and international diffusion of water conservation and availability technologies", *OECD Environment Working Papers*, OECD Publishing, Paris, forthcoming.

Egli, F. et al. (2015), "Identifying and inducing breakthrough inventions: An application related to climate change mitigation", *OECD Science, Technology and Industry Working Papers*, OECD Publishing, Paris, forthcoming.

Ellerman, D. (1998), "Note on the seemingly indefinite extension of power plant lives: A panel contribution", *The Energy Journal*, Vol. 19, No. 2, pp. 129-132, www.jstor.org/stable/41322782.

EC (2014), *BUILD UP Skills: An Initiative to Boost the Energy Skills of Europe's Building Workforce*, Executive Agency for Small and Medium-sized Enterprises, Publications Office of the European Union, Brussels, February, available at: http://ec.europa.eu/energy/intelligent/files/build_up_skills_publication.pdf.

Expert Group on Future Skills Needs (2010), "Future skills needs of enterprise within the green economy in Ireland", November 2010, Expert Group on Future Skills Needs Secretariat, Dublin, available at: www.skillsireland.ie/media/egfsn101129-green_skills_report.pdf.

Galgoczi, B. (forthcoming), "The long and winding road from black to green: Decades of structural change in the Ruhr region", *International Journal of Labour Research.*

Genon, G. and E. Brizio (2008), "Perspectives and limits for cement kilns as a destination for RDF", *Waste Management*, No. 28, pp. 2 375-2 385, http://dx.doi.org/10.1016/j.wasman.2007.10.022, cited in IPCC (2014).

Government of Ireland (2013), "Progress report on growth and employment in the green economy in Ireland", Department of Jobs, Enterprise and Innovation, Dublin, December, available at: www.djei.ie/publications/enterprise/2013/Green_Economy_Progress_Report_2013.pdf.

GTZ/Holcim (2006), "Guidelines on co-processing waste materials in cement production", Deutsche Gesellschaft Für Tchnische Zusammenarbeit GmbH, Eschborn, available at: www.holcim.fr/holcimcms/uploads/CORP/SummaryGuidelines2_COPROCEM.pdf, cited in IPCC (2014).

Hand, A. (2007), "Technology options for the cement industry with the use of alternative fuels", Alf-Cemind Workshop, Athens, 16 May, KHD Humboldt Wedag GmbH, available at: http://alf-cemind.com/docs/presentations/KHD%20presentation.pdf, cited in IPCC (2014).

Hasanbeigi, A. et al. (2012), *International Best Practices for Processing and Co-Processing Municipal Solid Waste and Sewage Sludge in the Cement Industry*, Lawrence Berkeley National Laboratory, Environmental Energy Technologies Division, Energy Analysis and Environmental Impacts Department, China Energy Group, LBNL-5581E, July.

Haščič, I. et al. (2015), "Public interventions and private climate finance flows: Empirical evidence from renewable energy financing", *OECD Environment Working Papers*, No. 80, OECD Publishing, Paris, http://dx.doi.org/10.1787/5js6b1r9lfd4-en.

Heyes, A. (2009), "Is environmental regulation bad for competition? A survey", *Journal of Regulatory Economics*, Vol. 36, No. 1, pp. 1-28, http://dx.doi.org/10.1007/s11149-009-9099-y.

IEA (2015), "Technology roadmaps", www.iea.org/roadmaps (accessed on 25 January 2015).

IEA (2013; 2014), CHP/DHC Country Scorecards: Japan (2013), Korea (2013), India (2014), United States (2014), The IEA CHP and DHC Collaborative, International Energy Agency, Paris, available at: www.iea.org/chp/countryscorecards.

IEA (2013), *Tracking Clean Energy Progress 2013*, IEA Input to the Clean Energy Ministerial, OECD/IEA, Paris, available at: www.iea.org/publications/tcep_web.pdf.

IEA (2012), *Energy Technology Perspectives 2012: Pathways to a Clean Energy System*, OECD/IEA, Paris, available at: www.iea.org/publications/freepublications/publication/industry2009.pdf.

IEA (2009), *Energy Technology Transitions for Industry: Strategies for the Next Industrial Revolution*, OECD/IEA, Paris, available at: www.iea.org/publications/freepublications/publication/industry2009.pdf.

IEA/WBCSD (2009), *Cement Technology Roadmap 2009: Carbon Emissions Reductions up to 2050*, International Energy Agency/World Business Council for Sustainable Development, www.wbcsd.org/Pages/EDocument/EDocumentDetails.aspx?ID=11423&NoSearchContextKey=true.

Johnstone, N. and I. Haščič (2012), "Increasing the penetration of intermittent renewable energy: Innovation in energy storage and grid management", in: OECD, *Energy and Climate Policy: Bending the Technological Trajectory*, OECD Publishing, Paris, http://dx.doi.org/10.1787/9789264174573-7-en.

Johnstone, N. et al. (2010), "Environmental policy design characteristics and technological innovation: Evidence from patent data", *OECD Environment Working Papers*, No. 16, OECD Publishing, Paris, http://dx.doi.org/10.1787/5kmjstwtqwhd-en.

Karstensen, K.H. (2008), "Formation, release and control of dioxins in cement kilns: A review", *Chemosphere*, Vol. 70, pp. 543-560, cited in Hasanbeigi et al. (2012).

Karstensen, K.H. (2007), "A literature review on co-processing of alternative fuels and raw materials and hazardous wastes in cement kilns", Department for Environmental Affairs and Tourism, Republic of South Africa, Pretoria, South Africa, available at: www.aitec-ambiente.org/Portals/2/docs/pubblici/Documenti/Raccolta%20bibliografica/Coprocessing%20literature%20review%202007.pdf, cited in IPCC (2014).

Kowalczyk, I. (2009), Testimony before the Senate Committee on Energy and Natural Resources, 7 May 2009, available at: www.ieca-us.com/wp-content/uploads/CHPOralTestimonyMay72009Senatehearing1.pdf.

Lerner, J. (2011), "Venture capital and innovation in energy", in: Henderson, R. and R. Newell (eds.), *Accelerating Innovation in Energy*, University of Chicago Press.

Levinson, A. (1999), "Grandfather regulations, new source bias, and state air toxics regulations", *Ecological Economics*, Vol. 28, No. 2, pp. 299-311, February, http://dx.doi.org/10.1016/S0921-8009(98)00045-7.

Miranda, G. and G. Larcombe (2012b), "Enabling local green growth: Addressing climate change effects on employment and local development", *OECD Local Economic and Employment Development (LEED) Working Papers*, No. 2012/01, OECD Publishing, Paris, http://dx.doi.org/10.1787/5k9h2q92t2r7-en.

Nelson, R.A. (1993), "Differential environmental regulation: Effects on electric utility capital turnover and emissions", *The Review of Economics and Statistics*, Vol. 75, No. 2, pp. 368-373, www.jstor.org/stable/i336976.

NRTEE (2012), *Framing the Future: Embracing the Low-carbon Economy*, Climate Prosperity Report 06, National Round Table on the Environment and the Economy,

Ottawa, Ontario, available at: http://publications.gc.ca/collections/collection_2012/trnee-nrtee/En133-40-6-2012-eng.pdf.

OECD (forthcoming), *The OECD Innovation Strategy Report*, OECD Publishing, Paris, forthcoming.

OECD (2015a), "The future of productivity", DSTI/IND(2015)1, OECD, Paris.

OECD (2015b), *Anticipating and Responding to Changing Skill Needs*, OECD Publishing, Paris.

OECD (2015c), *Investing in Youth Tunisia: Strengthening the Employability of Youth During the Transition to a Green Economy*, OECD Publishing, Paris, http://dx.doi.org/10.1787/9789264226470-en.

OECD (2014a), *OECD Science, Technology and Industry Outlook 2014*, OECD Publishing, Paris, http://dx.doi.org/10.1787/sti_outlook-2014-en.

OECD (2014b), "Identifying and inducing breakthrough inventions: An application related to climate change mitigation", DSTI/IND(2014)15, OECD, Paris.

OECD (2014c), "Measuring environmental innovation using patent data", ENV/EPOC/WPEI(2014)6, OECD, Paris.

OECD (2013a), "Nanotechnology for green innovation", *OECD Science, Technology and Industry Policy Papers*, No. 5, OECD Publishing, Paris, http://dx.doi.org/10.1787/5k450q9j8p8q-en.

OECD (2013b), "Biotechnology for the environment in the future: Science, technology and policy", *OECD Science, Technology and Industry Policy Papers*, No. 3, OECD Publishing, Paris, http://dx.doi.org/10.1787/5k4840hqhp7j-en.

OECD (2013c), *OECD Science, Technology and Industry Scoreboard 2013: Innovation for Growth*, OECD Publishing, Paris, http://dx.doi.org/10.1787/sti_scoreboard-2013-en.

OECD (2013d), "Business enterprise R-D expenditure by size class and by source of funds", *Research and Development Database*, **Error! Hyperlink reference not valid.**http://stats.oecd.org/Index.aspx?DataSetCode=BERD_SIZE (accessed March 2015).

OECD (2012a), *Energy and Climate Policy: Bending the Technological Trajectory*, OECD Studies on Environmental Innovation, OECD Publishing, Paris, http://dx.doi.org/10.1787/9789264174573-en.

OECD (2012b), "The jobs potential of a shift towards a low-carbon economy", *OECD Green Growth Papers*, No. 2012/01, OECD Publishing, Paris, http://dx.doi.org/10.1787/5k9h3630320v-en.

OECD (2012c), "ICT skills and employment: New competences and jobs for a greener and smarter economy", *OECD Digital Economy Papers*, No. 198, OECD Publishing, Paris, http://dx.doi.org/10.1787/5k994f3prlr5-en.

OECD (2011a), *Demand-side Innovation Policies*, OECD Publishing, Paris, http://dx.doi.org/10.1787/9789264098886-en.

OECD (2011b), *Invention and Transfer of Environmental Technologies*, OECD Studies on Environmental Innovation, OECD Publishing, Paris, http://dx.doi.org/10.1787/9789264115620-en.

OECD (2010), *The OECD Innovation Strategy: Getting a Head Start on Tomorrow*, OECD Publishing, Paris, http://dx.doi.org/10.1787/9789264083479-en.

OECD (2003), *The Environmental Performance of Public Procurement: Issues of Policy Coherence*, OECD Publishing, Paris, http://dx.doi.org/10.1787/9789264101562-en.

OECD/Cedefop (2014), *Greener Skills and Jobs*, OECD Green Growth Studies, OECD Publishing, Paris, http://dx.doi.org/10.1787/9789264208704-en.

Poirier, J. et al. (2015), "The benefits of international co-authorship in scientific papers: The case of wind energy technologies", *OECD Environment Working Papers*, No. 81, OECD Publishing, Paris, http://dx.doi.org/10.1787/5js69ld9w9nv-en.

Pôle Emploi (2011a), "Les emplois de la croissance verte, Enquête auprès des employeurs" (in French), *Repères et Analyses*, Vol. 20.

Pôle Emploi (2011b), "Les emplois de la croissance verte, Enquête auprès des demandeurs d'emploi" (in French), *Repères et Analyses*, Vol. 29.

Riesz, J. et al. (2013), "Payments for closure. Should direct action include payments for closure for high emission coal-fired power plants?", *CEEM Working Paper*, October, University of South Wales, Sydney, Australia, available at: http://ceem.unsw.edu.au/sites/default/files/documents/Working%20paper%20-%20Payments%20for%20Closure%20-%202013-10-07a.pdf.

Rovira, J. et al. (2011). "Use of sewage sludge as secondary fuel in a cement plant: Human health risks", *Environment International*, Vol. 37, cited in Hasanbeigi et al. (2012).

Salmi, O. et al. (2011), "Governing the interplay between industrial ecosystems and environmental regulation", *Journal of Industrial Ecology*, Vol. 16, No. 1, pp. 119-128, February, http://dx.doi.org/10.1111/j.1530-9290.2011.00403.x.

Sathaye, J. et al. (2011), "Bottom-up representation of industrial energy efficiency technologies in integrated assessment models for the cement sector", Ernest Orlando Lawrence Berkeley National Laboratory, Berkeley, California, available at: www.escholarship.org/uc/item/0xt5x113, cited in IPCC (2014).

Schorcht, F. et al. (2013), *Best Available Techniques (BAT) Reference Document for the Production of Cement, Lime and Magnesium Oxide*, Industrial Emissions Directive 2010/75/EU, JRC Reference Reports, European Commission, Joint Research Centre, Institute for Prospective Technological Studies, Publications Office of the European Union, Luxembourg, http://dx.doi.org/10.2788/12850.

SciColl (2013), *Scientific Collections International*, www.scicoll.org.

US EPA (2008), in conjunction with the US Department of Transportation and the US Department of Energy, Study on Increasing the Usage of Recovered Mineral Components in Federally Funded Projects Involving Procurement of Cement or Concrete to Address the Safe, Accountable, Flexible, Efficient Transportation Equity Act: A Legacy for Users, Report to Congress, EPA530-R-08-007, United States Environmental Protection Agency, Washington, DC, 3 June, available at: www.epa.gov/osw/conserve/tools/cpg/products/cement2.htm.

Wall Street Journal (2014), "Tesla Motors offers open licences to its patents – CEO Musk says approach will expand the market for electric cars", *The Wall Street Journal*, 12 June 2014, www.wsj.com/articles/tesla-motors-says-it-will-allow-others-to-use-its-patents-1402594375.

Warda, J. (2001), "Measuring the value of R&D tax treatment in OECD countries", *STI Review No. 27: Special Issue on New Science and Technology Indicators*, OECD Publishing, Paris, available at: www.oecd.org/sti/37124998.pdf.

WBCSD (2005), *Guidelines for the Selection and Use of Fuels and Raw Materials in the Cement Manufacturing Process*, World Business Council for Sustainable Development. www.wbcsdcement.org/index.php/publications, cited in IPCC (2014).

Zabaniotou, A. and C. Theofilou (2008), "Green energy at cement kiln in Cyprus: Use of sewage sludge as a conventional fuel substitute", *Renewable and Sustainable Energy Reviews*, Vol. 12, Issue 2, pp. 531-541, February, http://dx.doi.org/10.1016/j.rser.2006.07.017 cited in Hasanbeigi et al. (2012).

Chapter 5

Removing international trade barriers

Although international trade contributes directly to GHG emissions, increased trade can help to achieve economic goals in a GHG-efficient manner, provided that GHG emissions are correctly priced everywhere. Given that emissions are not universally priced, this chapter examines where policies related to trade may be in misalignment with climate change objectives. While concluding that the multilateral agreements of the World Trade Organization do not generally prevent governments from pursuing strong domestic climate policy, the chapter does identify potential misalignments. These include import tariffs on environmental goods, barriers to trade in services and domestic policies designed to support local low-carbon industry but which are restrictive of international trade and therefore potentially counter-productive. The chapter concludes by looking at pricing policies for the machinery of international trade, aviation and shipping fuel, as well as resilience of the global trade system.

Key messages

Trade itself is not the climate villain. If greenhouse gas (GHG) externalities were correctly priced globally, increased trade would lead to a more GHG-efficient global economy. However, not all GHG externalities are correctly priced. Given the importance of trade for shifting goods (and their embodied GHG emissions) internationally, it is important to assess how trade is likely to affect global GHG emissions – directly and indirectly – and where policy misalignments could lead to higher levels.

The international trade rules agreed through the World Trade Organization (WTO) do not generally impede governments from pursuing ambitious climate policies, even though they were not designed with the global climate challenge in mind. Nevertheless, other policies related to international trade should be designed to avoid misalignments with climate objectives.

How do tariff-based trade barriers affect the low-carbon transition? In many countries, particularly outside the OECD, import tariffs still exist for environmental goods, including goods important for climate change mitigation. Reducing these tariffs would help increase market access and promote the global transition to a low-carbon economy. Recent negotiations by a group of WTO members seeking an Environmental Goods Agreement are a promising development.

How can trade in services enhance the transition? The growth of global value chains has made international deployment of services a key part of modern trade, yet barriers to services trade still exist. Services are important for the low-carbon transition because more efficient services sectors contribute to improved productivity throughout the economy, often leading to lower energy and emissions intensity. Also, goods important for climate change mitigation are often new to markets and require highly skilled personnel to install, operate and maintain them. This means that widespread diffusion of such technologies, particularly in developing countries, could be particularly sensitive to barriers to trade in services. Removal of barriers such as restrictions on firm ownership and temporary movement of professionals could therefore be beneficial for the transition.

Are domestic support measures conducive to international trade? Many countries are promoting greener growth through the stimulation or creation of domestic industries manufacturing low-carbon power generating equipment. If unduly restrictive of international trade, however, these measures can undermine overall investment and uptake of the technology. One example is the recent prevalence of local-content requirements in the wind and solar energy sectors, aimed at supporting local mid-stream manufacturers. Such measures can increase overall costs due to the global nature of renewable energy value chains. Support measures should be designed to be respectful of international trade so as not to hinder international investment.

Are policies for aviation and maritime fuel aligned with climate objectives? GHG emissions from international merchandise transport may increase by 290% by 2050, with shipping and aviation accounting for more than 40% of freight emissions. Yet the international nature of these industries means that fuel destined for international shipping and aviation is often not covered by countries' core climate policies. Progress is being made in agreeing GHG-reduction measures at the multilateral level, however. For example, the International Maritime Organization has agreed energy efficiency standards and in 2016 the International Civil Aviation Organization will consider adopting a global market-based mechanism for pricing emissions from aviation. Pursuing multilateral efforts in these two domains is likely to be the most cost-effective means of ensuring full alignment with international climate change objectives.

Greenhouse gas emissions and international trade

International trade exerts an important influence on global efforts to tackle climate change. The interconnectedness of the global economy through trade means that countries' core climate policies do not operate in isolation. Short-term costs imposed by climate policies could lead to "carbon leakage" in cases where imports of carbon-intensive goods increase in response to more stringent mitigation efforts (see Chapter 1). Regardless of the effects of climate policies, international trade means that emissions generated in the production of exported goods (or intermediate products) will essentially be "consumed" in another country where the final good is purchased, presenting challenges for emissions accounting (Box 5.1). Trade is also an important driver of economic growth through the diffusion of technologies and know-how, including some that are important for the low-carbon transition. This chapter considers how policies relating to international trade could in some cases be misaligned with efforts towards the low-carbon transition.

Trade itself is not the climate villain. International trade does of course have direct emissions implications due to GHG emissions from transport (as well as other direct environmental impacts such as invasive alien species in containers and ballast water). However, when the life-cycle emissions of goods are taken into account, overseas production may still have lower emissions despite the international transport – how a product is produced is often more important than where it is produced. Further, the principle of free trade and comparative advantage suggests that over the long term, free and fair trade should lead to a more efficient outcome for the same level of economic output, assuming that climate-related externalities are correctly priced everywhere. In 2050, feeding 9 billion people all striving for wealthier lifestyles will be less resource-intensive with free trade than it would be without it, again assuming that GHG externalities are correctly priced.

However, all GHG emissions are not yet correctly priced. This means that it is important to assess how international trade is likely to affect global GHG emissions, and where policy misalignments could lead to higher GHG emissions. The environmental impacts of trade have often been framed in terms of their scale, composition and technique effects (Grossman and Krueger, 1993; Copeland and Taylor, 2003). When applied to GHG emissions, the scale effect refers to changes in emissions due to the increased economic activity from trade, including increased transport – which usually leads to increased emissions. The composition effect refers to changes in a country's emissions profile as relative prices and resource allocation between sectors adjust in response to international trade. As trade increases, some sectors will expand and others will contract in line with a country's comparative advantage, which could lead to either an increase or decrease in its overall emissions intensity, all else constant. The technique effect refers to improvements in emissions intensity due to innovation in the way goods and services are produced, such as through the diffusion of lower carbon goods and services.

Policy settings can influence how trade, through these three effects, influences GHG emissions. This chapter begins by looking for misalignments within international trade agreements and trade rules themselves. It then focuses on where domestic policies – including those intended to foster green growth – may be hindering the diffusion of low-carbon goods through international trade, influencing the technique effect. This chapter also assesses policies relating to international transport of goods, given their importance for influencing the direct emissions resulting from international trade. Finally, the role of policy in improving the resilience of the trading system in the face of physical climate impacts is briefly considered.

Box 5.1. Traded emissions: Calculating emissions based on production and consumption

A comparison of countries' production and consumption emissions can be visualised using data from the OECD's input-output tables combined with IEA data on CO_2 emissions.

Intellectually it might appear more appropriate to consider consumption-based emissions when assessing countries' efforts to reduce GHG emissions. If perfect information were available, it would be interesting to determine how a global carbon budget could be carved up based on the real emissions influenced by the consumption in each country. This would in theory remove any concerns about "carbon leakage" and allow each country to take responsibility for the emissions its economic activity really generates. In practice, at least two issues need to be considered. First, even though it can be claimed that a country is responsible for the emissions along global production chains generated by its economic activity, that country's possibilities to influence the emissions intensity abroad are limited. This is where an international agreement on territorial emissions continues to play an important role. Second, all GHG data are far from perfect, and agreeing on methods for measuring and comparing consumption-based emissions remains challenging (Lenzen et al., 2013; Nakano et al., 2009; Peters et al., 2011). Nevertheless, estimates such as those presented in Figure 5.1 provide a useful illustration of the importance of international trade for GHG emissions allocation. The data are similar to those presented for net export and import by region in the Intergovernmental Panel on Climate Change's (IPCC) Fifth Assessment Report (Agrawala et al., 2014).

Figure 5.1. **Production-based and consumption-based CO_2 emissions for selected countries**

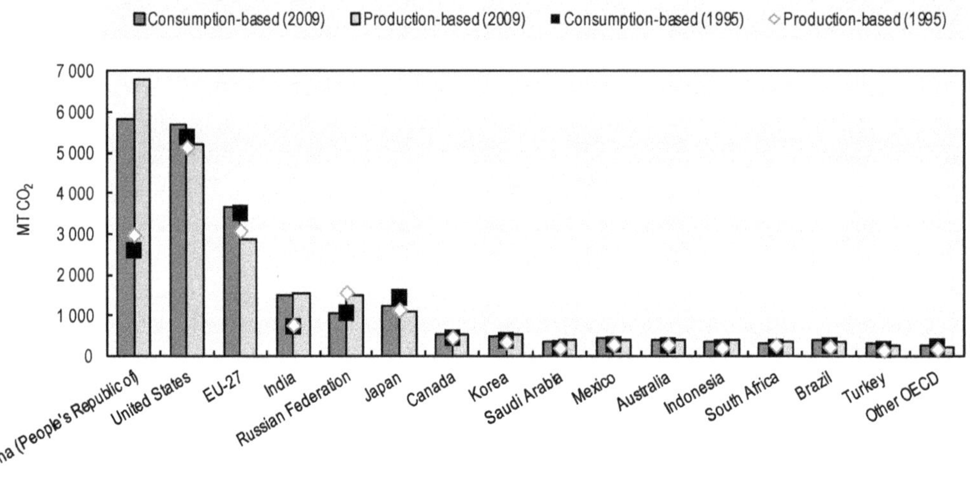

Source: Authors, based on OECD input-output tables and IEA CO_2 emissions data. See www.oecd.org/sti/inputoutput/co2.

Potential misalignments with international trade rules

The international trade regime includes rules agreed multilaterally under the WTO, rules agreed bilaterally or plurilaterally through regional trade agreements (RTAs), and jurisprudence from prior disputes relating to trade rules. Taken as a whole, does the trade regime act to restrict governments' ability to pursue ambitious climate policies? The following sections suggest that, in general, the trade regime is not in itself misaligned with climate objectives.

Multilateral agreements under the World Trade Organization

The WTO's primary agreement governing goods trade, the General Agreement on Tariffs and Trade 1994 (GATT), does not in itself prevent countries from pursuing climate policies. The GATT lays out the core principles for free trade. Key among these are the principles of non-discrimination between "like products" from different trading partners (most-favoured nation treatment) and between "like products" of foreign and domestic origin (national treatment). The question of whether products that differ only in the way they are produced – such as differences in GHG emissions during production – should be considered "like products" has been extensively debated by commentators and in ongoing WTO case law.

However, the GATT also allows for countries to justify policies on environmental (and other) grounds through Article XX, even if the measures partly violate one or more of the core principles.[1] Although the exemptions do not specifically mention climate change (the text dates from 1947), there is no clear evidence that the GATT in itself has acted to discourage countries from pursuing policies relating to climate change. In the few instances that WTO case law has tested whether climate change is an appropriate reason for justification under Article XX, opinions have generally been favourable (Tran, 2010).

One key and as yet untested question is the potential WTO conformity of border carbon adjustments that some countries may consider implementing as an economic disincentive targeted at controlling carbon leakage or limiting competitiveness concerns associated with mitigation policies (discussed in Chapter 1). No such measures have yet been introduced by any government, though examples have been considered in draft legislation and extensive academic literature on the issue exists (for a review see Condon and Ignaciuk, 2013). If such measures were to be introduced in the future and were subsequently challenged through the WTO, the interpretation of what constitutes a "like product" and the applicability of Article XX exceptions would again be important. But in general the WTO system does not explicitly forbid border carbon adjustments, whether in the form of taxes or through the purchase of emissions permits at the border.

Several of the more specific WTO agreements are also relevant to policies and measures targeting climate change objectives. One particular example is the Agreement on Subsidies and Countervailing Measures (SCM). Subsidies for the deployment of low-emitting technology have been one of the few policy tools readily available for governments seeking to take fast action on the low-carbon transition, given the barriers often faced when seeking to implement carbon pricing systems (described elsewhere in this report).

The SCM divides domestic subsidies into those prohibited for WTO members (i.e. contingent on export performance or on the use of domestic over-imported goods) and those that are "actionable" (not contingent on exports but potentially injurious to other WTO members, or contingent on domestic content). Although some observers have suggested revising the SCM's list of "non-actionable" subsidies to facilitate countries' efforts in implementing environmental subsidies (Charnovitz, 2014; Rubini, 2012), the prevalence of such measures already introduced by governments in recent years indicates that the SCM as it stands does not prevent action on "green" subsidies. Further, the SCM actually provides a means to ensure environmental subsidies are not trade-distorting or do not constitute disguised barriers to investment by foreign firms. Subsidies with these characteristics are potentially counterproductive for the low-carbon transition as they raise the costs of low-carbon technologies. These are further discussed below in the section on green industrial policy.

In general, the WTO Dispute Settlement Mechanism has allowed for jurisprudence to build up on an as-needed basis, with the application of trade rules to particular cases being clarified through emerging case law, including for measures related to climate change. In the case of subsidies, the dispute settlement process can lead to authorised unilateral trade remedies adopted by WTO members. Remedies such as anti-dumping and countervailing duties are legitimate, WTO-sanctioned responses to injuriously dumped or subsidised imports.[2] Recently, unilateral remedies have been applied in two directions within the same low-carbon industry. For example, the United States first imposed anti-dumping and countervailing duties on finished solar panels from the People's Republic of China (hereafter "China"). In response, China imposed similar measures on polysilicon precursors from the United States. The result of this escalation is reduced overall trade and increased costs in the supply chain (see review of studies in OECD, 2015). Although policy options for de-escalating trade remedies exist,[3] the costs incurred all across value chains and the uncertainty created for investors reinforce the importance of ensuring that domestic subsidies are designed in accordance with WTO principles, including the SCM.

Regional trade agreements

Outside of the WTO, governments have for many years pursued bilateral or plurilateral trade and investment agreements, often with the aim of creating closer ties with trade partners or moving towards deeper regional economic integration. Increasingly, these RTAs include specific environmental provisions (or environmental side agreements) which can be used to encourage more stringent environmental action (OECD, 2007; George, 2014). For example, provisions can include agreements to not weaken environmental laws in order to seek increased incoming international investment, and agreements to ensure that judicial enforcement capacity is available (e.g. the Peru-US and CAFTA-DR-US agreements; see US GAO, 2014 for a review). The effectiveness of these provisions depends on their degree of ambition, the extent to which they are binding on the parties, the stringency of their enforcement, and the nature and extent of co-operation between or among the parties to implement the provisions.

More recent RTAs aim to tackle behind-the-border barriers to trade in a more profound way than the WTO's Agreement on Technical Barriers to Trade (TBT). As well as chapters related to environment or sustainable development, these RTAs tend to include provisions on regulatory co-operation aiming to streamline regulations to reduce the cost of doing business internationally. Although this co-operation may cover environmental regulations, including those relevant to climate change mitigation, co-operation does not impede each party's sovereign right to regulate. Concerns have also been raised that investor-protection clauses, if included in RTAs where all parties have robust domestic investor protection laws, could be detrimental to the development of climate change policy measures. However, investor protection clauses have been used for many years and no conclusive evidence of this effect has been documented (Australian Productivity Commission, 2010; Tietje et al., 2014; BIAC, 2015).

Environmental goods trade liberalisation

Increased trade in environmental goods can help to mitigate environmental problems while also supporting economic growth – a manifestation of the "technique effect" described above. Most OECD countries have, over time, reduced their import tariffs for environmental goods, including those relevant to climate change mitigation. However, formal tariff-based trade barriers still exist for environmental goods, in particular outside the OECD area, with the result that the diffusion of some technologies important for addressing GHG emissions is hindered and costs in those countries are higher than they should be. Non-tariff barriers (NTBs) also hinder environmental goods dissemination sometimes to a larger extent than tariff barriers. Addressing NTBs as a source of policy misalignment could be explored further.

These barriers to trade can have a disproportionate effect on emerging technologies, including climate mitigation goods. With the rise of global value chains, intermediate products may cross borders many times before final assembly, meaning that even low tariffs can be amplified to a significant cost increase of the final product (OECD, 2013). Recent OECD work also showed that, in the case of some environmental goods, import tariffs can be detrimental to a country's exports, thereby hurting domestic industry (Sauvage, 2014). Furthermore, some low-carbon goods can face trade barriers because they are more capital-intensive than their fossil fuel counterparts, and import tariffs on fossil fuels are generally lower than on capital equipment.

The prospect of a multilateral agreement at the WTO with commitments on environmental goods tariffs has been discussed many times since 2001, so far with little progress in formal negotiations (Steenblik, 2005; Sauvage, 2014). Progress has been made outside of the WTO on a plurilateral basis. The Asia-Pacific Economic Cooperation (APEC) countries took a leading role in environmental goods trade by agreeing on the APEC List of Environmental Goods and committing to reduce applied tariff rates of the listed products to 5% or less by the end of 2015. In 2014, a group of WTO members, including OECD and non-OECD countries (among them China), commenced new plurilateral negotiations towards an Environmental Goods Agreement that is likely to include goods that are important for climate change mitigation (or are components thereof). If concluded successfully, such an agreement could potentially be formalised under the WTO in due course. Technical challenges remain, including reaching agreement on which goods should be considered for tariff liberalisation, given that many goods also have clearly non-environmental uses and are not separately identified in the Harmonized System (HS), the international classification and coding system used to track international trade (Steenblik, 2005; Sauvage, 2014).

Misalignments arising through domestic policies related to trade

Within the framework of the international trade regime, the trade effects of some domestic policies can have an important bearing on their effectiveness to support the low-carbon transition. These policies are examined in this section.

"Local-content requirements" for renewable energy

As part of their recovery from the financial crisis, many countries have implemented various forms of industrial policy, albeit often under different names (Evenett et al., 2009; Warwick, 2013). Where such policies directly support emissions-intensive investment, such as new subsidy arrangements favouring fossil fuels, misalignments with the low-carbon transition may exist that also distort international markets. These are covered in other chapters of this report.

A number of these newly introduced policies aim to promote green growth through the stimulation or creation of domestic industries manufacturing low-carbon power generation equipment. This trend has been referred to as the rise of "green industrial policy" (e.g. Wu and Salzman, 2014; Rodrik, 2013). Such measures may initially appear to be beneficial for the low-carbon transition. But various analyses have highlighted that if the measures are designed to be overly restrictive of international trade, they are likely to lead to higher prices for both domestic and international suppliers, with the overall effect of hindering uptake of low-carbon electricity-generating technologies.

Box 5.2 considers the specific and highly visible example of local-content requirements (LCRs) for renewable energy. These can be considered a policy misalignment for the low-carbon transition because they can raise the overall costs of downstream activities (e.g. installation). New OECD work indicates that LCRs have hindered both competitiveness and international investment in solar photovoltaics (PV) and wind energy. The increasingly globalised nature of value chains for wind and solar technology means that intermediate products cross borders many times. LCRs are usually intended to support mid-stream manufacturers, and the resulting market distortions can increase costs for actors further down the value chain. If these actors are in the same country, the policy may have a net negative effect for the domestic sector it is trying to support. Overall, such policies are likely to raise costs all across the production chain (Bahar et al., 2013; OECD, 2015).

The risk of higher overall costs also exists in relation to other trade-impacting "behind the border" measures in the same sectors. These include measures with more direct trade implications (such as local-equity requirements and export quotas) and those that deter international investment and therefore lead to overall less efficient supply chains (e.g. national standards that favour domestic producers or more informal measures that favour local enterprises over foreign ones). The prevalence of these measures – and the WTO disputes associated with them – highlights the need for policy makers to better align and take a more holistic approach to trade and investment policies in order to provide effective policy support to achieve the low-carbon transition.

Barriers to trade in services

Over time, the global importance of trade in services has risen significantly. Global value chains and highly streamlined international logistics networks have made international deployment of services a key part of modern trade. The value created by services as intermediate inputs now represents over 30% of the total value added in manufactured goods. The international trade regime addresses services trade through the General Agreement on Trade in Services (GATS), agreed in 1994. However, negotiations on specific liberalisation commitments under the agreement have faltered over time and many barriers to trade in services remain in the form of domestic regulations.[4] Some of these are important for the low-carbon transition.

Trade in services is important for climate change mitigation in a number of ways. In general, more efficient services sectors contribute to improving productivity and enhancing competitiveness across the whole economy – in manufacturing as well as in services sectors themselves (OECD, 2014). Greater productivity will often lead to lower energy use and emissions intensity. Also, as economies become ever more interconnected through value chains, a trend towards "servicification" can be identified, with companies increasingly turning to provision of services attached to the delivery of goods. For example, a jet engine manufacturer is more likely to lease its engines to airlines, and an industrial turbine manufacturer is more likely to lease its turbine. This usually leads to better maintenance

Box 5.2. Local-content requirements in renewable energy markets

Local-content requirements (LCRs) have increasingly been used to support the development of renewable energy. OECD research shows that LCRs linked to wind and solar PV have been planned or implemented in at least 21 countries, including 16 OECD countries, mostly since 2009. LCRs are typically imposed as a precondition for access to financial support schemes such as feed-in tariff (FiT) programmes or as part of eligibility requirements in renewable energy public tenders. Some countries have also designed LCRs as eligibility criteria for direct financial transfers such as subsidised loans and loan guarantees from government agencies and national development banks, such as in Brazil. In some cases, different LCR ratios are used depending on the technology used in downstream installations, such as India (OECD, 2015; OECD et al., 2013; Bahar et al., 2013).

To highlight the effects of LCRs on international investment, new OECD research based on empirical analysis indicates that while FiT policies play an important role in attracting international investment in solar PV and wind energy, LCRs have a detrimental effect on global international investment flows in these sectors and hinder the effectiveness of FiT policies when attached to them. The estimated detrimental effect of LCRs is slightly stronger when both domestic and international investments are considered. This indicates that LCRs do not have positive impacts on domestic investment flows (OECD, 2015). At the same time, recent OECD Computable General Equilibrium (CGE) modelling has shown an array of expected negative impacts of LCRs on trade across different sectors (Stone et al., 2015).

The rise of LCRs for renewable energy has led to at least five WTO disputes since 2010, highlighting the importance that governments place on new renewable energy industries. In one example, the Canadian province of Ontario was found to be in breach of GATT (General Agreement on Tariffs and Trade) and TRIMS (Agreement on Trade-Related Investment Measures) commitments, though the FiT subsidy itself was not found to be in breach of the SCM (Agreement on Subsidies and Countervailing Measures).

Ontario's FiT programme (introduced in 2010) required developers of wind and solar installations benefiting from the FiT to have a certain percentage of project costs incurred in Ontario. Japan launched a complaint at the WTO and a similar complaint was later lodged by the European Union. The claimants argued that the "Minimum Required Domestic Content Level" adopted under the FiT Programme unfairly discriminated against foreign renewable-energy equipment and therefore placed Canada in violation of: 1) the national treatment obligation under Article III:4 of the GATT 1994; 2) the prohibition set out in Article 2.1 of the TRIMs Agreement on the application of any trade-related investment measures that are inconsistent with Article III of the GATT 1994; 3) the prohibition on import substitution subsidies prescribed in Articles 3.1(b) and 3.2 of the SCM.

In 2013 the WTO's Appellate Body confirmed that the LCR was inconsistent with international trade rules. The Appellate Body supported the panel's conclusions that LCRs accorded preferential treatment to products made in Ontario by requiring the purchase or use of products from domestic sources, placing Canada in breach of its national treatment obligation under GATT Article III and TRIMS Article II. Following the ruling, Canada proceeded to bring its measure into conformity with the recommendations of the WTO's Dispute Settlement Body (DSB). On 5 June 2014, Canada informed the DSB that the Government of Ontario was: 1) no longer subjecting large, renewable-based electricity procurements to domestic requirements; 2) had significantly lowered its domestic-content requirements for small and micro-FiT procurement relating to solar- and wind-powered electricity installations under the FiT Programme.

Source for final two paragraphs: Based on www.wto.org/english/tratop_e/dispu_e/cases_e/ds412_e.htm.

and performance of the equipment, resulting in lower fuel use and lower emissions. It is also likely to lead to better overall utilisation rates of physical capital, thereby contributing to a more energy efficient economy. But to be effective, this "servicification" of the economy requires smooth international trade in services (Swedish National Board of Trade, 2014).

Concerning specific technologies important for climate change, such as renewable energy, the deployment of technology is dependent on a wide range of services, many of which are imported and are not necessarily strictly environmental in nature – particularly in the context of developing countries. Business services, telecommunications services, and construction and related engineering services figure prominently (Steenblik and Geloso Grosso, 2011). Low-carbon goods tend to be newer, high-tech goods requiring highly skilled personnel to install, operate and maintain. Training of local users can also be important. Overall this means that widespread diffusion of such technologies, particularly in developing countries, is likely to be more affected by barriers to services trade than "conventional" more highly emitting goods.[5] Finally, services that are traditionally considered to be "environmental services", such as pollution remediation, may also be important for climate mitigation.

Tracking and understanding trade in services is difficult due to data constraints. Recently, the OECD developed the Services Trade Restrictiveness Index (STRI) to shed light on barriers to services trade across different sectors and countries (Box 5.3). Although it has not developed an index specifically for environmental services, those for other service industries highlight where some countries could do more to remove barriers to services trade that would help their efforts toward the low-carbon transition. Of the four modes of services trade identified in the GATS,[6] Steenblik and Geloso Grosso (2011) document examples of all four being relevant to climate change. These range from consulting services for energy efficiency (Mode 1); to ecotourism services consumed abroad (Mode 2); to the establishment of foreign subsidiaries to manage low-carbon projects (Mode 3); to temporary movement of personnel such as to carry out wind turbine repairs (Mode 4). The Swedish National Board of Trade (2014) identified a list of services indispensable to trade in environmental goods; these also cover all four modes but with Mode 3 (commercial presence) and Mode 4 (natural movement of persons) predominating.

Barriers to cross-border trade in electricity

As described in Chapter 7, the uptake of renewable energy is an important means for mitigating CO_2 emissions from the electricity sector, but the variable output of some renewable energy technologies can introduce instability into electricity systems. The risk of instability increases with higher shares of intermittent power sources connected to the electrical grid. Chapter 7 delves into detail on electricity market structures that may be poorly aligned with the transition to a low-carbon economy. Barriers to cross-border trade in electricity appear to be another potential misalignment. An examination of European electricity markets confirms the importance of cross-border electricity trade in increasing the effective capacity factor of intermittent plants in the context of a growing share of intermittent renewables in the power sector (Bahar and Sauvage, 2013).

Box 5.3. Services Trade Restrictiveness Index (STRI)

Since 2014, the OECD has been tracking barriers to services trade across countries and sectors through the Services Trade Restrictiveness Index (STRI). The STRI contains a regulatory database of laws and regulations in existence today, and composite indices that quantify identified restrictions across five standard categories, with values between zero and one. A score of zero corresponds to complete openness to trade and investment, while being completely closed to foreign services providers yields a score of one.

The STRI provides a unique diagnostic tool, generating a picture of services restrictiveness at the national level and by sector, covering 18 sectors in 40 countries. It allows benchmarking for individual countries and relative to global best practice, and enables countries to quickly see where the outlier restrictions are and where potential bottlenecks exist.

For the first time, comprehensive and comparable information is available for policy makers to scope out reform options and assess their likely effects; for trade negotiators to clarify those restrictions that most impede trade; and for businesses to understand entry requirements for foreign markets. The knock-on consequences for downstream users of these services are demonstrable. The STRI in combination with the OECD-WTO TiVA-GVC database are powerful tools for further analysis of regulatory spillovers in global value chains and the interdependence between sectors in an interconnected and increasingly digital world.

Figure 5.2 shows an example of STRI data for engineering services, a key service area relevant to climate change technology. Engineering services are labour-intensive, particularly at the high-skill level. Therefore, measures categorised under "Restrictions to movement of people" have the strongest impact in the restrictiveness levels for these services. The other policy category that affects the degree of restrictiveness in engineering services relates to "Restrictions on foreign entry". Some countries maintain ownership restrictions on the basis of qualifications and licensing, at times coupled with residency and licensing requirements for board members and managers of engineering firms. More open services markets improve competitiveness and productivity both in the services sectors in question and downstream industries using services as inputs. Engineering services underpin the infrastructural development of the economy and the smooth functioning of essential public services. Hence, promoting the cost-effectiveness and quality of these services can represent a source of economic growth and create significant spill-over effects.

Figure 5.2. **STRI* by policy area: Engineering services**

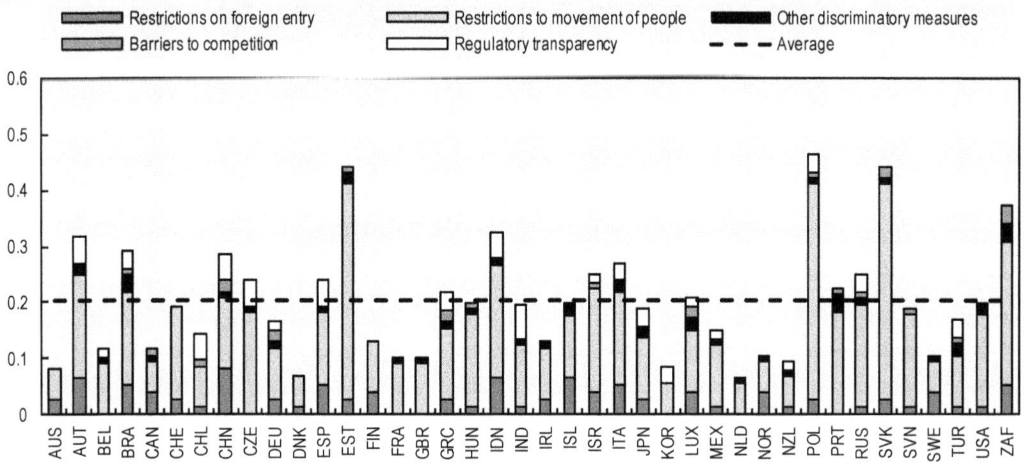

Note: *The STRI indices take values between zero and one, one being the most restrictive.

Source: OECD (2014), "Services Trade Restrictiveness Index: Policy brief", OECD, Paris, available at: www.oecd.org/tad/services-trade/STRI%20Policy%20Brief_ENG.pdf.

The primary constraint for cross-border trade is availability of physical infrastructure in the form of interconnectors joining national grids. However, even in the presence of sufficient interconnection capacity, other barriers may still exist. The existence of an efficient cross-border electricity trading regime takes on particular importance to address the variability problem. Addressing those regulatory and administrative measures that inhibit cross-border trade and the smooth operation of regional electricity markets would therefore help increase the potential for trade in electricity to facilitate the uptake of renewable energy (Bahar and Sauvage, 2013).

Fuelling international trade: Maritime shipping and aviation

Although not trade policies *per se*, policies relating to international freight transport are considered here because of their importance to both international trade and global GHG emissions. The movement of goods in international trade predominantly relies on fossil fuels, mainly fuels derived from petroleum. The impact that increased international trade will have on global GHG emissions trends – via the scale and direct effects described above – depends in part on the policy incentives in place internationally for freight transport.

Transport as a whole accounts for 23% of global CO_2 emissions from fuel combustion (30% in OECD countries). The majority of this is road transport (of which only about one-third is attributable to freight and international trade), with shipping and aviation making up about 3% and 2% respectively of total global CO_2 emissions from fuel combustion (OECD/ITFb, 2015). However, projections by the International Transport Forum (ITF) suggest that in the base case, GHG emissions from international merchandise transport may increase by 290% by 2050 (OECD/ITF, 2015a). Although the majority of international merchandise trade is still expected to travel by sea, road transport accounts for around 50% of total CO_2 emissions from international trade-related freight movements; the share is projected to increase to 56% by 2050.[7]

Shipping and aviation are for the most part international industries, governed by specific international conventions.[8] The United Nations has multilateral organisations devoted to both domains: the International Civil Aviation Organization (ICAO) and the International Maritime Organization (IMO). The international character of these industries makes it difficult for individual governments to take action on pricing or regulating GHG emissions for bunker fuels consumed on their territories or under their flag. However, some progress has been made through both international conventions, as described in the following sections.

Policies relating to emissions from international aviation

The multilateral Chicago Convention (1944) originally granted tax exemptions to fuel carried on-board planes arriving within the territory of an ICAO member. In 2000, the ICAO Council agreed to broaden the scope of the exemption to also cover fuel taken on-board prior to leaving a country's airspace, reflecting what many countries had already agreed in bilateral air services agreements (ICIS, 2005).

Countries originally sought international agreement on reciprocal exemption of aviation fuel from tax because the mobile nature of aviation may otherwise have resulted in tax competition across countries. If one country's tax rate was notably higher than others, it is possible that airlines would refuel in other countries with lower taxes, a practice known as "tankering". Depending on relative fuel costs, this could result in suboptimal fuel loads leading to overall increases in fuel use, and therefore emissions.

Despite these tax exemptions, progress has been made regionally and internationally on improving technical standards and pricing emissions. As a result of rising fuel costs over the past decade, airlines and states have invested in technologies, equipment and operations to lower the overall consumption of fuel. Through the ICAO, a CO_2 metric has been developed with a view to agreeing an international CO_2 standard for aircraft. National and regional technical initiatives also exist.[9]

In 2012, the EU sought to introduce carbon pricing for aviation by bringing aviation within the EU ETS and requiring all departing and incoming flights to surrender emissions permits (some of which are allocated free of charge to operators). The strong international reaction, including a questioning of the legality of the EU's move and threats of retaliatory trade measures aimed at EU aviation industries, led to a reduction in scope of the measure to cover only intra-EU flights, pending progress on an international market-based mechanism under the ICAO.

In 2013, ICAO members formally agreed to explore a multilateral market-based solution. The 2016 ICAO Assembly is due to take a decision on this proposal shortly after the important UNFCCC COP21 in December 2015. Should the decision on and implementation of the multilateral mechanism be subject to delays, governments seeking to take strong action on climate change may need to explore interim solutions bilaterally or on a plurilateral basis. The 2000 decision by the ICAO Council is a "recommended practice" and as such is not legally binding on parties and does not preclude members, individually or jointly, from introducing environmental taxes on aviation fuel.

Policies relating to emissions from international maritime transport

Maritime transport generates the lowest GHG emissions per tonne-km compared with the other modes of commercial transport, yet the vast quantity of goods transported by sea means that overall emissions have grown quickly (faster than road transport, against a 1990 baseline; IEA, 2014). However, according to the 3rd IMO GHG Study, global shipping emissions were relatively flat between 2007 and 2012, totalling 972 Mt CO_2-eq in 2012 (IMO, 2014).

On a multilateral basis, countries have made progress through the IMO to reach agreements on improving the fuel efficiency of ships and on agreeing progressive reductions on the maximum limits for air pollutants for ships, including SO_x and NO_x. In 2011, the IMO successfully amended MARPOL Annex VI when it introduced the Energy Efficiency Design Index, which established a mandatory efficiency standard for the design of new ships (measured in terms of GHG emissions per tonne-mile), and also the Ship Energy Efficiency Management Plan, an operational measure that establishes a mechanism to improve ships' energy efficiency. Together these measures could reduce shipping GHG emissions by 180 Mt annually by 2020 (Lloyds Register and DNV, 2011).

Air pollution controls can also have a GHG effect. Given that the main way to reduce air pollutants from ships is currently through purchasing higher quality fuel (distillate as opposed to residual fuels), air pollution controls have the effect of raising fuel costs, thereby creating an incentive to reduce GHG emissions in line with local air pollutants. Although exhaust filters (scrubbers) are increasingly used in maritime transport, stricter air pollution controls would also act as an incentive to reduce GHG emissions.

While no agreement has yet been reached to develop a market-based or other pricing mechanism for GHG emissions from ships, the IMO is working on the development of further technical and operational measures. The current focus is on the development of a global data collection system for GHG emissions from ships, but it is hoped that in the long term a global market-based measure to address GHG emissions from shipping will be developed.

Resilience of the modern trade system to climate change

Modern global value chains (GVCs) have become increasingly international, connected and reliant on domestic policies that are open to international trade and fair to international investors. Intermediate goods may cross borders many times in their journey from primary material to finished goods. Expedient movement of goods, machinery and people is essential to ensure that the global production machine has a sufficient supply of services and materials to keep it running smoothly.

Recent OECD work on global value chains (OECD, 2013) points out that increasingly the "just-in-time" nature of value chains makes them quite vulnerable to external shocks. The OECD defines global shocks as "rapid-onset events with severely disruptive consequences covering at least two continents" (OECD, 2011). Two recent examples are highlighted in OECD (2013), where major physical events in one part of the world caused measurable knock-on effects for global industries. One example, not climate-related, is the earthquake and tsunami in Japan in 2011, which had considerable knock-on effects on the global electronics and automotive industries. Another example is flooding in Thailand in 2012, which at its peak covered areas accounting for 45% of the world's manufacturing capacity of computer hard disk drives and led to global disruptions not only in the computer industry but also the automotive industry (OECD, 2013).

Climate-related events such as flooding and severe storms are likely to intensify due to climate change, thus increasing the systemic risk inherent in GVCs. Companies are already responding by complementing "just-in-time" with "just-in-case" contingency plans and seeking trade-offs between cost minimisation and security of supply. Companies are seeking to diversify risks geographically and between different suppliers, and there is some evidence of a trend towards "back-shoring" or "near-shoring" with GVCs being splintered into shorter chains. The OECD has helped countries understand their vulnerability to shocks via the TIVA database (OECD et al., 2013), and is helping governments to better understand GVC risks through the G20-OECD Framework for Disaster Risk Management and the *OECD Principles for Country Risk Management* (OECD, 2013).

When considering alignment issues in national strategies for climate change adaptation and resilience, it will be increasingly important to consider how each country's position and role in GVCs – and the national policies shaping the participation of firms in those value chains – could be developed to ensure resilience in the face of increasingly frequent and severe weather-related shocks.

Notes

1. If a policy measure related to climate change mitigation seeks exemption from goods trade rules as a necessary measure for the low-carbon transition, the measure must satisfy the content of one of the paragraphs of Article XX. In most environmental cases this means the measure must be "relating to the conservation of exhaustible natural resources" or be "necessary to protect human, animal or plant life or health". The measure seeking exemption must also satisfy the chapeau of the article – that is, not to constitute an "arbitrary or unjustifiable discrimination between countries where the same conditions prevail" or a "disguised restriction on international trade".

2. For countervailing duties, the implementing party must demonstrate that "specific" subsidies were provided that caused "injury" to the domestic complaining industry before countervailing duties can be imposed. Export subsidies and local content subsidies – which are generally prohibited – are deemed specific. For all other subsidies, the subsidy must be shown to be limited to a specific company or industry, or group of companies or industries. Subsidies that are not prohibited, are not specific or do not cause injury are permissible under WTO rules.

3. These include reductions in the level of the duty imposed (not seeking to counter the full value of the dumping), reducing the scope (e.g. to specific product or import value) or targeting only companies with a dominant anti-competitive market position (Wu and Salzman, 2014; Swedish National Board of Trade, 2013).

4. Progress is being made on a plurilateral basis, In 2013, a group of 23 WTO members started plurilateral negotiations on a specific Trade in Services Agreement (TISA) that follows GATS principles and aims to establish commitments between signatories in areas such as licensing, financial services, telecoms, e-commerce, maritime transport, and professionals moving abroad temporarily to provide services.

5. Exceptions do of course exist, such as technologies to convert coal to liquids and for extracting and refining oil sands, both of which involve higher life-cycle emissions of GHG than producing petroleum from many conventional wells.

6. Mode 1, cross-border trade (the supplier is not present in the country in which the service is supplied); Mode 2, consumption abroad (an individual travels to a foreign country where the service is supplied); Mode 3, commercial presence (a service is supplied through a subsidiary established in the host country); Mode 4, movement of natural persons (an individual travels abroad to supply a service in a host country or to work as an intra-corporate transfer under Mode 3).

7. Policies concerning road transport are covered in Chapters 3 and 8.

8. Domestic flights and shipping routes tend to be treated separately in domestic policies; these are not covered here.

9. See, for example, efforts through NextGen in the United States (www.faa.gov/nextgen) and SESAR in the European Union (www.sesarju.eu).

References

Agrawala, S. et al. (2014), "Regional development and cooperation", in: *Climate Change 2014: Mitigation of Climate Change*, Contribution of Working Group III to the Fifth Assessment Report of the Intergovernmental Panel on Climate Change, Cambridge University Press, Cambridge, United Kingdom and New York, New York, United States, available at: www.ipcc.ch/pdf/assessment-report/ar5/wg3/ipcc_wg3_ar5_full.pdf.

Australian Productivity Commission (2010), *Bilateral and Regional Trade Agreements*, Productivity Commission Research Report, Commonwealth of Australia, Melbourne, Australia, November, available at: www.pc.gov.au/__data/assets/pdf_file/0010/104203/trade-agreements-report.pdf.

Bahar, H. and J. Sauvage (2013), "Cross-border trade in electricity and the development of renewables-based electric power: Lessons from Europe", *OECD Trade and Environment Working Papers*, No. 2013/02, OECD Publishing, Paris, http://dx.doi.org/10.1787/5k4869cdwnzr-en.

Bahar, H. et al. (2013), "Domestic incentive measures for renewable energy with possible trade implications", *OECD Trade and Environment Working Papers*, No. 2013/01, OECD Publishing, Paris, http://dx.doi.org/10.1787/5k44srlksr6f-en.

BIAC (2015), "Investor-state dispute settlement: An indispensable element of investment protection", Business and Industry Advisory Committee to the OECD, available at: www.investmentpolicycentral.com/sites/g/files/g798796/f/201502/BIAC%20FIN%2015-01%20ISDS%20paper.pdf.

Charnovitz, S. (2014), "Green subsidies and the WTO", *World Bank Policy Research Working Paper*, No. 7060, World Bank Group, Climate Change Group, Washington, DC, October, available at: http://documents.worldbank.org/curated/en/2014/10/20290817/green-subsidies-wto.

Condon, M. and A. Ignaciuk (2013), "Border carbon adjustment and international trade: A literature review", *OECD Trade and Environment Working Papers*, No. 2013/06, OECD Publishing, Paris, http://dx.doi.org/10.1787/5k3xn25b386c-en.

Copeland, B. and M. Taylor (2003), *Trade and the Environment: Theory and Evidence*, Princeton University Press, Princeton and Oxford.

Evenett, S.J. et al. (2009), "Effective crisis response and openness: Implications for the trading system", in: Evenett, S.J. et al. (eds.), *Effective Crisis Response and Openness: Implications for the Trading System*, Centre for Economic Policy Research, London, pp. 249-262.

George, C. (2014), "Developments in regional trade agreements and the environment: 2013 update", *OECD Trade and Environment Working Papers*, No. 2014/01, OECD Publishing, Paris, http://dx.doi.org/10.1787/5jz0v4qcg9zw-en.

Grossman, G. and A. Krueger (1993), "Environmental impacts of a North American Free Trade Agreement", in: Garber, P.M. (ed.), *The US-Mexico Free Trade Agreement*, MIT Press, Cambridge, Massachusetts.

Howse, R. (2010), "Climate mitigation subsidies and the WTO legal framework: A policy analysis", International Institute for Sustainable Development, Winnipeg, Manitoba, Canada, May, available at: www.iisd.org/pdf/2009/bali_2_copenhagen_subsidies_legal.pdf.

ICIS (2005), "Tax flight: An investigation into the origins and developments of the exemption from various kinds of taxation of international aviation", *Working Paper I05-E001*, International Centre for Integrative Studies, Maastricht University, Maastricht, Netherlands, available at: www.thepep.org/ClearingHouse/docfiles/taxflight_final.pdf.

IEA (2014),

IMO (2014), "Reduction of GHG emissions from ships", *Third IMO GHG Study 2014 – Final Report*, International Maritime Organization, London, available at: www.iadc.org/wp-content/uploads/2014/02/MEPC-67-6-INF3-2014-Final-Report-complete.pdf.

Lenzen, M. et al. (2013), "Building eora: A global multi-regional input-output database at high country and sector resolution", *Economic Systems Research*, Vol. 25, No. 1, pp. 20-49.

Lloyds Register and DNV (2011), "Assessment of IMO mandated energy efficiency measures for international shipping: Estimated CO_2 emissions reduction from introduction of mandatory technical and operational energy efficiency measures for ships", Lloyd's Register and DNV, available at: www.imo.org/MediaCentre/HotTopics/GHG/Documents/REPORT%20ASSESSMENT%20OF%20IMO%20MANDATED%20ENERGY%20EFFICIENCY%20MEASURES%20FOR%20INTERNATIONAL%20SHIPPING.pdf.

Nakano, S. et al. (2009), "The measurement of CO_2 embodiments in international trade: Evidence from the Harmonised Input-Output and Bilateral Trade Database", *OECD Science, Technology and Industry Working Papers*, No. 2009/03, OECD Publishing, Paris, http://dx.doi.org/10.1787/227026518048.

OECD (2015), "Overcoming barriers to international investment in clean energy", COM/DAF/INV/ENV/EPOC(2014)1, OECD, Paris, forthcoming.

OECD (2014), "Services Trade Restrictiveness Index: Policy brief", OECD, Paris, available at: www.oecd.org/tad/services-trade/STRI%20Policy%20Brief_ENG.pdf.

OECD (2013), *Interconnected Economies: Benefiting from Global Value Chains*, OECD Publishing, Paris, http://dx.doi.org/10.1787/9789264189560-en.

OECD (2011), *Future Global Shocks: Improving Risk Governance*, OECD Publishing, Paris, http://dx.doi.org/10.1787/9789264114586-en.

OECD (2007), *Environment and Regional Trade Agreements*, OECD Publishing, Paris, http://dx.doi.org/10.1787/9789264006805-en.

OECD/ITF (2015), *ITF Transport Outlook 2015*, OECD Publishing, Paris, http://dx.doi.org/10.1787/9789282107782-en.

OECD/ITF (2010), "Reducing transport greenhouse gas emissions: Trends & data 2010", OECD/ITF, Paris, available at: www.internationaltransportforum.org/Pub/pdf/10GHGTrends.pdf.

OECD, WTO and UNCTAD (2013), "Implications of global value chains for trade, investment, development and jobs", paper prepared for the G20 Leaders Summit, Saint Petersburg (Russian Federation), September, available at: www.oecd.org/sti/ind/G20-Global-Value-Chains-2013.pdf.

Peters, G.P. et al. (2011), "Constructing an environmentally extended multi-regional input-output table using the GTAP Database", *Economic Systems Research*, Vol. 23, No. 2, pp. 131-152, http://dx.doi.org/10.1080/09535314.2011.563234.

Rodrik, D. (2013), "Green industrial policy", Grantham Research Institute project on "Green Growth and the New Industrial Revolution", available at: www.sss.ias.edu/files/pdfs/Rodrik/Research/Green-growth-and-industrial-policy.pdf.

Rubini, L. (2012), "Ain't wastin' time no more: Subsidies for renewable energy, the SCM Agreement, policy space, and law reform", *Journal of International Economic Law*, http://dx.doi.org/10.1093/jiel/jgs020.

Sauvage, J. (2014), "The stringency of environmental regulations and trade in environmental goods", *OECD Trade and Environment Working Papers*, No. 2014/03, OECD Publishing, Paris, http://dx.doi.org/10.1787/5jxrjn7xsnmq-en.

Steenblik, R. (2009), "Green growth, protectionism, and the crisis", in: Evenett, S.J. et al. (eds.), *Effective Crisis Response and Openness: Implications for the Trading System*, Centre for Economic Policy Research, London, pp. 249-262.

Steenblik, R. (2005), "Liberalising trade in 'environmental goods': Some practical considerations", *OECD Trade and Environment Working Papers*, No. 2005/05, OECD Publishing, Paris, http://dx.doi.org/10.1787/888676434604.

Steenblik, R. and M. Geloso Grosso (2011), "Trade in services related to climate change: An exploratory analysis", *OECD Trade and Environment Working Papers*, No. 2011/03, OECD Publishing, Paris, http://dx.doi.org/10.1787/5kgc5wtd9rzw-en.

Stone, S. et al. (2015), "Emerging policy issues: Localisation barriers to trade", *OECD Trade Policy Papers*, OECD Publishing, Paris, forthcoming.

Swedish National Board of Trade (2014), *Making Green Trade Happen: Environmental Goods and Indispensable Services*, Kommerskollegium, Stockholm, available at: www.kommers.se/Documents/dokumentarkiv/publikationer/2014/Making-Green-Trade-Happen_webb.pdf.

Swedish National Board of Trade (2013), *Effects on Trade and Competition of Abolishing Anti-Dumping Measures: The European Union Experience*, Kommerskollegium, Stockholm, available at: www.kommers.se/Documents/dokumentarkiv/publikationer/2013/rapporter/report-effects-on-trade_webb.pdf.

Tietje, C. et al. (2014), "The impact of investor-state-dispute settlement (ISDS) in the Transatlantic Trade and Investment Partnership", study prepared for the Minister for Foreign Trade and Development Co-operation, Netherlands, available at: www.rijksoverheid.nl/bestanden/documenten-en-publicaties/rapporten/2014/06/24/the-impact-of-investor-state-dispute-settlement-isds-in-the-ttip/the-impact-of-investor-state-dispute-settlement-isds-in-the-ttip.pdf.

Tran, C. (2010), "Using GATT, Art XX to justify climate change measures in claims under the WTO agreements", *Environmental and Planning Law Journal*, Vol. 27, available at: http://ssrn.com/abstract=1676105.

US GAO (2014), *Free Trade Agreements – Office of the U.S. Trade Representative Should Continue to Improve its Monitoring of Environmental Commitments*, Report to Congressional Requesters, November, GAO-15-161, United States Government Accountability Office, Washington, DC, available at: www.gao.gov/assets/670/666782.pdf.

Warwick, K. (2013), "Beyond industrial policy: Emerging issues and new trends", *OECD Science, Technology and Industry Policy Papers*, No. 2, OECD Publishing, Paris, http://dx.doi.org/10.1787/5k4869clw0xp-en.

Wu, M. and J. Salzman (2014), "The next generation of trade and environment conflicts: The rise of green industrial policy", *Northwestern University Law Review*, Vol. 108, pp. 401-474, http://scholarship.law.duke.edu/faculty_scholarship/3320.

Chapter 6

Diagnosing misalignments for a more resilient future

Climate change increases the risk of severe, pervasive and irreversible impacts such as species extinctions, threats to food security, changing patterns of weather-related mortality and rising sea levels. All countries are or will be affected; yet, many are not well adapted to this new reality. This chapter first reviews misalignments between existing policies and climate adaptation objectives, such as regulatory barriers in infrastructure financing, poorly designed planning policies and lack of pricing of natural resources. It then provides guidance on how countries could better manage the risk linked to climate change.

Key messages

Adapting to an already changing climate will also mean addressing policy misalignments. In the same way the economic fabric and regulatory environment have been marked by the convenient and pervasive use of fossil fuels, the climate experienced to date has partly locked societies and ecosystems in patterns that make them vulnerable to a changing climate. This century could witness a rate and scale of change unprecedented in human history. All countries will be affected, with the poor and socially marginalised being the hardest hit.

Misalignments with climate adaptation include: regulatory regimes for infrastructure that deter investment in resilience; planning policies that encourage development in vulnerable areas; and under-pricing of natural resources. Although there would be benefits from addressing these issues in any case, climate change strengthens the urgency of doing so. Many countries are now developing strategic, national approaches to adaptation to systematically address these barriers.

Adaptation will not remove all the risks from climate change. Managing the financial impact of remaining risks is essential for building resilience. There is also a need to ensure that risk-sharing and risk-transfer arrangements do not exclude the needs of the poor.

All countries will be affected by climate change

Adaptation has become an increasingly important component of the policy response to climate change, as the world is now committed to some degree of global warming irrespective of future emissions. The Intergovernmental Panel on Climate Change (IPCC) projects increases in global mean surface temperature of between 3.7°C and 4.8°C by 2100 relative to the second half of the 19th century in the absence of strengthened mitigation policies (IPCC, 2014). This would represent a pace and scale of climate change unprecedented in human history. The IPCC notes that higher warming increases the risk of "severe, pervasive and irreversible impacts". These include species extinctions, threats to food security, changing patterns of weather-related mortality and rising sea levels.

Yet, in many cases, countries are not well-adapted to current climate variability. This is illustrated by the toll of weather- and climate-related disasters. Munich RE, a re-insurance company, estimated that these events led to 19 000 deaths and economic losses of USD 125 billion in 2013 alone (Munich RE, 2014). In the last ten years, OECD and BRIC (Brazil, Russian Federation, India and China) countries have experienced an estimated USD 1.5 trillion in economic damages from disasters including storms or floods (OECD, 2014a). Table 6.1 summarises how each component of these total losses could be affected by climate change.

All countries will be affected by climate change, but they will not all be affected equally. Climate risks will depend upon of the extent of climate change, the location of people and assets, and the extent to which societies and ecosystems are able to adapt to a changing climate. The scale of the adaptation challenge is largest for developing countries, where resources and capacity to respond to climate risks are most clearly constrained. Urbanisation entailing the construction of infrastructure and assets is an integral part of the development process. However, depending on where they occur, the ensuing concentration of people and assets in high-risk locations can in some cases ultimately increase vulnerability to effects of climate change such as flooding (OECD, 2014a).

Figure 6.1. **Economic losses due to disasters in OECD and BRIC countries, 1980-2012**

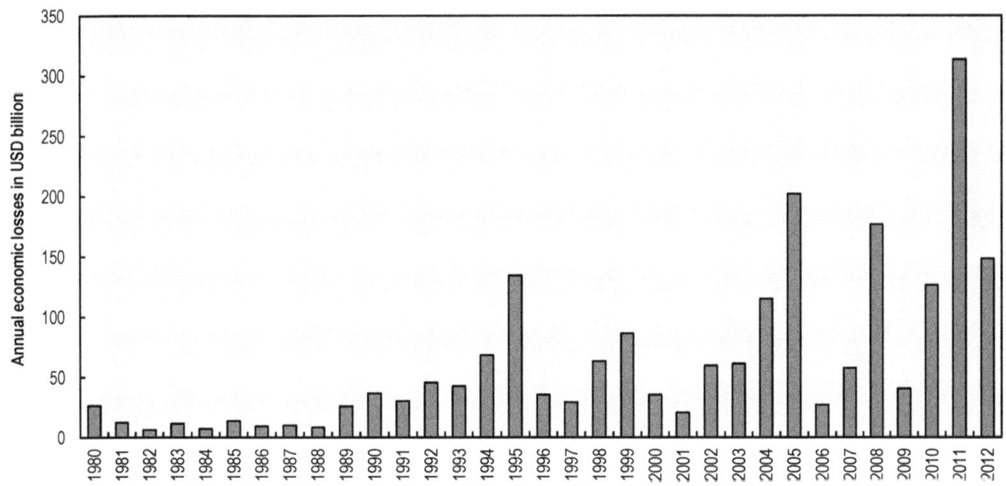

Source: OECD (2014b), *OECD Reviews of Risk Management Policies: Boosting Resilience through Innovative Risk Management*, OECD Publishing, Paris, http://dx.doi.org/10.1787/9789264209114-en.

Table 6.1. **Climate change projections of insured losses and/or insurance prices**

Hazard	Insurance line	Region	Projected changes in future time slices relative to current climate (spatial distribution and vulnerability of insured values assumed to be unchanged over time)
Winter storm	Homeowners' insurance	Europe	- Projected increases in mean annual loss ratio (e.g. Belgium, France, Germany, Ireland, Netherlands, Poland and the United Kingdom) lie in a range of 1-to-2 digit percentages before and around 2050, with larger increases at the end of the century. In southern Europe (e.g. Portugal and Spain) losses are expected to decrease. - Today's 20-year, 10-year and 5-year return periods will for all of Europe be roughly halved by the end of the century.
River flood, marine flood, flash flood from rainfall, melting snow	Property and business interruption insurances	Europe, North America	- Canada: losses from heavy precipitation in property and business interruption insurances in three city areas are projected to rise by 13% (2016-35), 20% (2046-65) and 30% (2081-00). - Germany: projected increases in mean annual insured flood loss are 84% (2011-40), 91% (2041-70) and 114% (2071-00). - The Netherlands: expected annual property loss caused by increasing river discharge and sea-level rise with an assumed flood insurance system is projected to lie 125% higher in 2040 relative to 2015 (with a 24 cm sea-level rise) and by 1 784% higher in 2100 (85 cm sea-level rise). - Norway: in three counties across southern Norway precipitation and snow melt insurance losses are expected to be higher by approximately 10-21% and 17-32% at the end of the century. - United Kingdom: projected increases in mean annual insured flood loss are 8% for a 2°C global mean temperature rise and 14% for a 4°C rise, with the 1-in-100-year loss higher by 18% and 30%, respectively.
Tropical cyclone	Foremost property insurance lines	Asia, North America	- China: projected increases of insured typhoon losses are 20% (for a 2°C Scenario) and 32% (for a 4°C Scenario), with the 1-in-100-year loss higher by 7% and 9%, respectively. -United States: Florida's hurricane wind insurance is projected to change by –20% to +5% (2020s) and –28% to +10% (2040s) (under the assumptions of strained reinsurance capacity).
Hailstorm	Homeowners' insurance, agricultural insurances	Europe	- Germany: projected increases in mean annual loss ratios from homeowners' insurance due to hail are 15% (2011-40) and 47% (2041-70). -The Netherlands: losses from outdoor farming insurance and greenhouse horticulture insurance are projected to increase by 25-29% and 116-134%, respectively, in a 1°C Scenario. For a 2°C Scenario, projected increases are 49-58% and 219- 269%, respectively.
Storms, pests, diseases	Paddy rice insurance	Asia	- Japan: paddy rice insurance payouts are projected to decrease by 13% by the 2070s, on the basis of changes in standard yield and yield loss.

Source: Adapted from IPCC (2014), *Climate Change 2014: Impacts, Adaptation, and Vulnerability*, Contribution of Working Group II to the Fifth Assessment Report of the Intergovernmental Panel on Climate Change, Field, C.B. et al. (eds.), Cambridge University Press, Cambridge, United Kingdom and New York, New York, United States.

OECD countries have much higher inherent adaptive capacity, but the consequences of extreme weather events have illustrated that they remain vulnerable. Long-time horizons and diffuse responsibility can prevent the allocation of finance and capacity development required to manage climate risk (Mullan et al., 2013). This is a particular issue at the local level, where responsibility for implementing adaptation actions frequently lies. Furthermore, in an interconnected world the effects of climate change do not stop at national borders: knock-on effects can be felt though trade links, financial markets and movements of people (OECD, 2011; 2014b).

Conversely, international links also represent an opportunity. Diversification of supply chains can aid systemic robustness to localised shocks. The transfer of ideas and innovation policy can help with the development and diffusion of climate-resilient technologies. International co-operation will be essential for overcoming the effects of climate change. Official development assistance (ODA) and other forms of development co-operation have increasingly mainstreamed climate change adaptation across their portfolios, although gaps remain.

Uncertainty represents an overarching challenge for an effective adaptation to climate change. There is reasonable confidence in likely global average temperature increases, but much less certainty about how precipitation patterns and intensity will change regionally, and about the incidence and severity of extreme weather events. This uncertainty about the climate is compounded with uncertainty about where, when and how climate events will affect economic, social and environmental systems. The greater the degree of climate change, the higher the probability of tipping points with potentially severe economic and social consequences.

Box 6.1. **Definitions of key terms**

Climate resilience is the capacity of individuals and social, economic or environmental systems to absorb and recover from climate-induced shocks, while adapting and transforming their structures and means of living in the face of long-term stresses, change and uncertainty (Mitchell, 2013; UNISDR, 2013).

Adaptation is "the process of adjustment to actual or expected climate and its effects to moderate harm or exploit beneficial opportunities" (IPCC, 2012).

Risks are "the potential for consequences due to climate change where something of value is at stake and where the outcome is uncertain, recogni[s]ing the diversity of values. Risk is often represented as probability of occurrence of hazardous events or trends multiplied by the impacts if these events or trends occur" (IPCC, 2014).

Disaster risk management is the systematic process of using administrative directives, organisations and operational skills and capacities to lessen the adverse impacts of hazards and the possibility of disaster. It involves activities and measures for prevention, mitigation and preparedness (UNISDR, 2009).

Source: Adapted from OECD (2014a), *Climate Resilience in Development Planning: Experiences in Colombia and Ethiopia*, OECD Publishing, Paris, http://dx.doi.org/10.1787/9789264209503-en.

Building resilience to climate impacts

Mitigation policies are primarily designed to solve a single market failure: the negative externality of greenhouse gas (GHG) emissions. The issue is more complex for building resilience to climate change, where there is no single metric or outcome to target. People and ecosystems will inevitably respond to a changing climate, but the form that this response takes will be affected by institutions, infrastructure and the information that they are faced with. Market failures, physical and cultural constraints, and policy misalignments can all act as barriers to efficient and equitable adaptation. Governments have an important role to play in overcoming these barriers to private sector adaptation, while also ensuring that public sector actions contribute to resilience (Cimato and Mullan, 2011).

Government action to facilitate adaptation can take a wide variety of forms, including providing information, strategic planning, regulatory and fiscal measures (de Bruin et al., 2009). However, broadly speaking, two policy areas will be crucial for adaptation: reducing risks and managing those that remain. These two policy areas are closely interconnected in practice. For example, the availability of insurance (a mechanism for managing residual risks) can affect people's incentives to reduce risks, while the extent of risk reduction affects the affordability of insurance (G20/OECD, 2012). Risk reduction measures (such as protective infrastructure) can provide protection from recurrent events and reduce insurance costs, while the cost efficiency of insurance tends to improve with decreasing frequency (OECD, 2014b).

Countries are increasingly taking action to systematically bring together risk reduction and improve their risk financing in the face of climate change. For example, national adaptation plans or strategies have been developed in 23 out of 34 OECD countries (OECD, 2015). Other countries are implementing strategic national approaches without developing specific plans. Meanwhile, 50 least developed countries laid the foundation for a co-ordinated approach to adaptation through the development of national adaptation plans of action. Following the Cancun agreement in 2011, developing countries are being supported to develop national, strategic approaches to climate change.

Risk reduction

Climate change will result in a diverse and extensive range of risks, and the scope of measures to reduce those risks is correspondingly broad. As well as "hard" infrastructure measures, such as the construction of coastal and flood defences, "soft" measures are also necessary to manage risks (de Bruin et al., 2009). These measures include information provision, regulatory reforms and institutional capacity building. Even though there can be direct benefits to the people implementing these measures, market failures and policy misalignments can prevent them from being implemented. Table 6.2 illustrates the range of potential impacts of climate change and their corresponding risk reduction measures.

Only a small set of policies are solely motivated by the desire to adapt to climate change: these tend to focus on institutional reforms and the provision of evidence of climate risks (Mullan et al., 2013). The implementation of specific adaptation responses is increasingly being mainstreamed, i.e. systematically integrated into the policy development process (de Bruin et al., 2009; OECD, 2015). This enables climate risk to be considered in concert with other trends, facilitating a co-ordinated and effective response.

Although the concept of mainstreaming risk reduction is straightforward, implementation remains at an early stage in many areas (OECD, 2015). Examples of areas where reforms may be needed are discussed in more detail in the thematic chapters, but examples include:

- planning policies that encourage, or implicitly subsidise, development in high-risk areas
- water management plans that fail to consider the potential for ecosystem-based approaches to water supply and flood risk management

- infrastructure regulations that can lead to underinvestment in resilience.

Table 6.2. **A range of potential climate risks and response measures**

Sector	Examples of risks	Risk reduction measures
Agriculture	- Reduced crop yields/quality - Heat stress	- Alter crop species - Alter timing - Irrigation - Plant breeding
Coastal areas	- Damage to infrastructure and buildings from sea-level rise	- Relocate vulnerable infrastructure - Sea wall construction - Beach enrichment
Water	- Drought - Ecological damage - Flooding	- Invest in water efficiency - Divert/store more water - Flood zoning
Health	- Reduction in winter mortality/increase in summer mortality - Changing incidence of disease	- Prepare for extreme weather - Control disease carriers - Increase surveillance of infectious diseases
Energy	- Changing patterns of demand - Supply disruption	- New cooling capacity - Changes in insulation - New building codes

Source: Adapted from de Bruin, K. et al. (2009), "Economic aspects of adaptation to climate change: Integrated assessment modelling of adaptation costs and benefits", *OECD Environment Working Papers*, No. 6, OECD Publishing, Paris, http://dx.doi.org/10.1787/225282538105.

As is the case with GHG emissions for mitigation, mispricing of natural resources and subsidies can distort market adaptations. For example, they can soften the incentive for farmers to adjust practices, or encourage the use of inputs (such as freshwater) that may become scarcer in the future.

In many cases, policies to address existing challenges can also support longer term resilience, for example by increasing the efficiency of water use by households. More generally, and given uncertainties over climate impacts, OECD guidance recommends that policy be designed to achieve flexibility, give rise to co-benefits or perform well under a wide range of possible climate outcomes (OECD, 2013; Agrawala and Fankhauser, 2008).

Managing residual risks

It is not cost-effective, nor feasible, to reduce the risks from climate change to zero. The management of the risks that remain after risk reduction measures is crucial for building climate resilience. Some risks will have to be accepted, while others can benefit from financial management. The G20/OECD methodological framework, *Disaster Risk Assessment and Risk Financing* (2012) framework publication on risk assessment and risk financing identifies how the financial management of residual risks can be improved. Adverse climate shocks have direct impacts on households and businesses, but the effects are also felt by national governments and throughout the financial system. Putting measures in place *ex ante* can increase efficiency, provide clarity of incentives and help to dampen the effects of negative shocks. These efforts can be implemented domestically or through international co-operation. Table 6.3 summarises the available tools for *ex ante* risk financing and lists the *ex post* responses that would have to be used to address impacts where *ex ante* measures are unavailable or insufficient. In countries with under-developed financial systems, these *ex post* responses can magnify the negative impact of negative shocks. For example, the forced sale of livestock in the face of drought will erode households' future earnings potential.

Table 6.3. **Summary of risk management, risk pooling and risk transfer approaches**

Level of risk transfer	*Ex ante* measures	*Ex post* responses
National	- Insurance pools - Contingent credit - Catastrophe bonds	- Budget reallocation - Debt financing or borrowing - Humanitarian relief
Domestic financial institution	- Diversification - Capital reserves - Re-insurance - Transferring risks to capital markets through securitisation	- Insolvency/bail-out - Reduction in credit
Household	- Savings and credit - Informal risk sharing - Micro-insurance	- Selling productive assets - Reducing food consumption - Borrowing money (often at very high interest rates) - Migrating - Taking on additional work

Source: Revised from G20/OECD (2012), "Disaster risk assessment and risk financing: A G20/OECD methodological framework", OECD, Paris, available at: www.oecd.org/gov/risk/G20disasterriskmanagement.pdf.

Policies towards risk financing and insurance are not always conducive to strengthening resilience, however. Common challenges include:

- an inability of traditional mechanisms to reach the poorest and most vulnerable
- lack of incentives to reduce risk – e.g. premiums not linked to exposure
- unclear and *ad hoc* policies regarding the possibility of *ex post* indemnification.

As an example, experience with the US National Flood Insurance Program, in which reforms to encourage risk reduction ended up being reversed due to political opposition, illustrates the barriers to reform even if the defects of current arrangements are well understood (Box 6.2).

There will be consequences of climate change that remain after the implementation of risk-reduction and risk-transfer measures. The consequences of these residual impacts will depend upon underlying determinants of resilience such as gross domestic product (GDP) per capita, quality of institutions and social cohesion.

Economic resilience has historically been strong for OECD countries, but this could change as the scale of losses increase. In contrast, negative climate shocks in developing countries can have lasting negative impacts on growth and poverty (OECD, 2014a). Additional challenges include:

- impacts on some local and regional economies will be more pronounced than at the national level, particularly where there is a lack of risk-sharing between levels of government
- the effects of impacts could spread out in unpredictable ways due to global value chains, e.g. the Bangkok flooding in 2011
- if extreme weather events become more severe or frequent, historical experience may not hold in the future.

Box 6.2. Reform of the US National Flood Insurance Program

The US National Flood Insurance Program (NFIP) was created in 1968 to help property owners to financially protect themselves against floods associated with hurricanes, tropical storms, heavy rains and other climate events. Participation in the programme is mandatory for all properties with mortgages from federally regulated or insured lenders located in high flood-risk areas (defined as a one-in-four chance of flooding during a 30-year mortgage). Participation is not required for properties in moderate- to low-risk areas, but they account for nearly 25% of flood insurance claims and one-third of Federal Disaster Assistance for flooding.

A 2010 report from the Government Accountability Office identified a number of design features that constrain the fiscal soundness of the programme and impede the efficient management of flood risk. They include: statutory limits on rate increases, the inability to reject high-risk applicants, the mismatch between NFIP premiums and the real flood risks for almost a quarter of property owners, the use of "grandfathered" rates for some properties not taking into account reassessments of flood risk, and the inability of the programme to deny coverage to repetitive loss properties accounting for 25-30% of insurance claims while only making up 1% of policies. It is estimated that a series of disasters including Hurricanes Sandy and Katrina have contributed to a negative balance of close to USD 24 billion.

To address these challenges, in 2012 the US Congress passed the Biggert-Waters Flood Insurance Reform Act (BW-12). The objective of the reform was to target the fiscal soundness of the programme, promote more efficient risk management and assess future changes to flood risk based on the best available scientific evidence. This resulted in annual premium rate increases for policy holders of up to 20% (twice the previous limit) based on calculations of an "average historical loss year", including catastrophic loss years. Subsidies were also phased out for a number of properties, severe repetitive loss properties in particular. Further, flood insurance rate maps were updated to include, *inter alia*, data and information on changes in sea levels, precipitation and hurricane intensity.

Political opposition against BW-12 ultimately led to the passage of the Homeowner Flood Insurance Affordability Act in March 2014, which repeals and modifies certain provisions of BW-12. For example, instead of an immediate premium increase to full-risk rates, the Affordability Act requires (with a few exceptions) that the increase be gradual, but no less than 5% and no more than 18% annually. Further, the act reinstates the use of "grandfathered" rates that enable policy holders of new properties to benefit the first year from premium rates offered to properties located outside the Special Hazard Area. Policy holders in high-risk areas required to pay their full-risk rate under BW-12 are also entitled to refunds, while policy holders that face an 18% premium increase may be entitled to refunds. Despite these amendments, the objective of the Affordability Act is to make the NFIP self-sufficient by gradually moving towards actuarial rates. All policies for primary residences will also be subjected to a USD 25 surcharge, while all other policies include a USD 250 surcharge.

Sources: GAO (2010), *National Flood Insurance Program. Continued Actions Needed to Address Financial and Operational Issues, Testimony Before the Subcommittee on Housing and Community Opportunity, Committee on Financial Services*, House of Representatives, United States Government Accountability Office, Washington, DC; FEMA (2014), Homeowner Flood Insurance Affordability Act, US Federal Emergency Management Agency, Washington, DC; OECD (2013), *Water and Climate Change Adaptation: Policies to Navigate Uncharted Waters*, OECD Studies on Water, OECD Publishing, Paris, http://dx.doi.org/10.1787/9789264200449-en.

Social resilience (the resilience of people and communities), is less easily quantified, but is also a challenge for all countries. Part of the difficulty is that many social impacts cannot be readily transferred or shared, so there is no intermediate layer between resilience and risk reduction – e.g. the mental health impacts of flooding, injuries and deaths from heat waves. These are also harder to quantify, but the impacts can be significant, including an erosion of trust in government, looting and anti-social behaviour. The trend changes caused by climate change, such as loss of ecosystem services due to shifting climatic zones, may also weaken resilience, and the effects of this will be felt when extreme events occur.

Addressing the pervasive policy misalignments affecting adaptation is an essential step for building resilience. Substantive progress has been made in areas with clear accountabilities and visible costs of maintaining the status quo, e.g. management of flood risks (OECD, 2013). However, as the reforms to the US flood insurance programme have demonstrated, even in these areas the political economy can be challenging. The situation is more difficult where misalignments cross sectoral boundaries or are likely to arise mainly in the longer term, such as development in vulnerable areas. In addressing these areas, strategic approaches to adaptation (e.g. the development of national adaptation plans) can help to identify where there is the greatest need for reform. To be effective, these should be accompanied by the development of robust mechanisms to monitor and assess progress in building resilience, though most countries are at an early stage in defining and measuring success (OECD, 2015).

References

Agrawala, S. and S. Fankhauser (eds.) (2008), *Economic Aspects of Adaptation to Climate Change: Costs, Benefits and Policy Instruments*, OECD Publishing, Paris, http://dx.doi.org/10.1787/9789264046214-en.

Cimato, F. and M. Mullan (2011), "The role of government in adapting to climate change", *Defra Economics Working Paper 1*, Department of Environment, Food and Rural Affairs, London.

de Bruin, K. et al. (2009), "Economic aspects of adaptation to climate change: Integrated assessment modelling of adaptation costs and benefits", *OECD Environment Working Papers*, No. 6, OECD Publishing, Paris, http://dx.doi.org/10.1787/225282538105.

FEMA (2014), *Homeowner Flood Insurance Affordability Act*, US Federal Emergency Management Agency, Washington, DC.

G20/OECD (2012), "Disaster risk assessment and risk financing: A G20/OECD methodological framework", OECD, Paris, available at: www.oecd.org/gov/risk/G20disasterriskmanagement.pdf.

GAO (2010), *National Flood Insurance Program. Continued Actions Needed to Address Financial and Operational Issues*, Testimony Before the Subcommittee on Housing and Community Opportunity, Committee on Financial Services, House of Representatives, United States Government Accountability Office, Washington, DC.

IPCC (2014), *Climate Change 2014: Synthesis Report*, Intergovernmental Panel on Climate Change, available at: www.ipcc.ch/pdf/assessment-report/ar5/syr/SYR_AR5_FINAL_full.pdf.

IPCC (2012), "Summary for policymakers", in: Field, C.B. et al. (eds.), *Managing the Risks of Extreme Events and Disasters to Advance Climate Change Adaptation*, A Special

Report of Working Groups I and II of the Intergovernmental Panel on Climate Change, Cambridge University Press, Cambridge, United Kingdom and New York, New York, United States, pp. 1-19.

Mechler, R. et al. (2014), "Managing unnatural disaster risk from climate extremes", *Nature Climate Change*, Vol. 4, No. 4, pp. 235-237, April, www.nature.com/nclimate/journal/v4/n4/full/nclimate2137.html.

Mitchell, A. (2013), "Risk and resilience: From good idea to good practice", *OECD Development Co-operation Working Papers*, No. 13, OECD Publishing, Paris, http://dx.doi.org/10.1787/5k3ttg4cxcbp-en.

Mullan, M. et al. (2013), "National adaptation planning: Lessons from OECD countries", *OECD Environment Working Papers*, No. 54, OECD Publishing, Paris, http://dx.doi.org/10.1787/5k483jpfpsq1-en.

Munich RE (2014), *Loss Events Worldwide 2013*, NatCatSERVICE, www.munichre.com/site/touch-naturalhazards/get/documents_E1193185391/mr/assetpool.shared/Documents/5_Touch/Natural%20Hazards/NatCatService/Annual%20Statistics/2013/MunichRe-Natcatservice-Naturaldisaster2013-Perc-Distrib-Event-by-Type.pdf

OECD (2015), *Economics of Adaptation*, OECD Publishing, Paris, forthcoming.

OECD (2014a), *Climate Resilience in Development Planning: Experiences in Colombia and Ethiopia*, OECD Publishing, Paris, http://dx.doi.org/10.1787/9789264209503-en.

OECD (2014b), *OECD Reviews of Risk Management Policies: Boosting Resilience through Innovative Risk Governance*, OECD Publishing, Paris, http://dx.doi.org/10.1787/9789264209114-en.

OECD (2013), *Water and Climate Change Adaptation: Policies to Navigate Uncharted Waters*, OECD Studies on Water, OECD Publishing, Paris, http://dx.doi.org/10.1787/9789264200449-en.

OECD (2011), *Towards Green Growth*, OECD Green Growth Studies, OECD Publishing, Paris, http://dx.doi.org/10.1787/9789264111318-en.

OECD (2008), *Economic Aspects of Adaptation to Climate Change: Costs, Benefits and Policy Instruments*, OECD Publishing, Paris, http://dx.doi.org/10.1787/9789264046214-en.

UNISDR (2013), *From Shared Risk to Shared Value: The Business Case for Disaster Risk Reduction,* United Nations Global Assessment Report on Disaster Risk Reduction, United Nations Office for Disaster Risk Reduction, Geneva, available at: www.unisdr.org/we/inform/publications/33013.

UNISDR (2009), *2009 UNISDR Terminology on Disaster Risk Reduction*, United Nations Office for Disaster Risk Reduction, Geneva, available at: www.unisdr.org/we/inform/publications/7817.

Part II

Aligning policies in specific activities

Chapter 7

Reframing investment signals and incentives in electricity

Electricity generation and uses are critical to the decarbonisation of energy systems. In some countries, power generators compete on wholesale electricity markets that optimise the near-term power supply. This chapter describes how such design may limit the market's ability to guide future low-carbon investments, including in presence of a price on CO_2. New arrangements are needed to lock-in investment in capital-intensive low-carbon technologies, while reflecting their specific costs and benefits for electricity system. The incentives in regulated electricity systems are also not always aligned with the decarbonisation of power generation. Beyond generation, there are examples of regulatory barriers that could be addressed to facilitate climate-friendly innovations, such as demand-side response and electricity storage. The chapter also touches on the impacts of climate change on energy systems.

Key messages

Electricity, an increasingly widespread and convenient energy carrier, is essential to the decarbonisation of the energy sector and of the economy as a whole. It can substitute for fossil fuels in many end-uses. Low-carbon substitutes are also available in generation, although national endowments and technology choices mean that not all options are available in all countries. Global CO_2 emissions from electricity (and heat) still account for 25% of total greenhouse gases (GHGs) and are on the rise.

The regulatory framework for electricity systems is critical, as it eventually determines the investment context, cost and reliability of the system. The question is whether current regulatory frameworks and markets can adequately reflect the CO_2 constraint in investment and operational decisions.

Should electricity market design be revamped for the low-carbon transition? Current designs of wholesale electricity markets in many OECD countries are not strategically aligned with the low-carbon transition. They do not deliver the long-term price signal that investment in high capital cost, low-carbon technologies such as hydro-electricity, nuclear power, wind turbines, solar technologies and geothermal installations or fossil fuelled-plants fitted with carbon capture and storage (CCS) would require.

A standard energy-only wholesale electricity market would require a high CO_2 price, periods of very high electricity prices and risks of rolling brown-outs before investors would unlock financing in low-carbon technologies. The associated risks would lead to higher capital costs and higher-than-necessary electricity costs. The result would be ongoing investment in CO_2-emitting generation.

Ensuring competitive and timely investment in low-carbon capacity and competition in electricity supply among existing plants requires new market arrangements as well as a robust CO_2 price. At present no general model exists, but options are being tested in OECD and non-OECD countries.

Low-carbon technologies play different roles in electricity systems. Hydropower and nuclear plants, biomass-fuelled plants and CCS-fitted plants are dispatchable – i.e. can generate when required. Variable sources such as wind and solar PV are not fully dispatchable. More flexible power systems are thus needed for a high penetration of variable sources. The costs of system transformation vary widely in different countries and should be allocated fairly and transparently to avoid distortions.

Potential misalignments also exist which prevent a broader engagement of the demand side of electricity markets and the deployment of storage. Both constitute areas of dynamic development and could enhance the flexibility of electricity systems in the future.

Will regulated systems do better in the transition? Regulated electricity systems (e.g. with no liberalised market structure) also face challenges such as fair grid and market access for new low-carbon sources and their adequate remuneration. Further, the experience of decarbonisation in wholesale electricity markets should provide useful lessons for governments seeking to introduce competition in regulated electricity systems. The adoption cost of low-carbon technologies will be strongly influenced by the nature of contracts awarded through competition.

Are policies helping the resilience of the energy sector? Energy systems and infrastructures are exposed to extreme weather events as well as gradual changes driven by climate change. This is a serious consideration for fossil-fuelled and low-carbon power plants. Energy companies, users and policy makers are only starting to explore the policy aspects of climate impacts.

Electricity in decarbonisation

The energy path to decarbonisation

The latest energy and climate scenarios of the International Energy Agency (IEA) give several indications on the features of global energy supply and demand apt to respond to climate change (IEA, 2014a). Two major differences in the global energy picture in 2050 appear when comparing a business-as-usual scenario – a 6°C Scenario (6DS) – with a scenario compatible with the 2°C goal (2DS):

- A much more efficient use of energy: the same energy services (i.e. mobility, heat and all the specific uses of electricity) would be provided with 27% less primary energy.
- A reduced share of fossil fuels in primary energy supply, with biomass and waste, hydro-electricity, nuclear power and other renewables accounting for roughly 60% of the total. In volume, non-fossil energy would nearly quadruple between 2011 and 2050.

As presented in Chapter 1, putting the energy system on a path consistent with the 2°C objective requires a combination of the core climate policy instruments mentioned earlier: a growing price on CO_2 emissions, targeted measures to improve energy efficiency and support measures for the deployment of low-carbon technologies. Alternative scenarios were also run by the IEA assuming a higher penetration of renewables (making up for lower contributions from nuclear power and carbon capture and storage, or CCS), a higher electrification of transport and a higher penetration of electricity in end-uses in buildings.

The central role of electricity in decarbonisation

A view of a decarbonised electricity system in 2050

Electricity is called on to play a prominent role in the global decarbonisation of energy, as shown in Figures 7.1 and 7.2. This phenomenon would be common to all world regions,[1,2] with several trends at play in this particular scenario (IEA, 2014a):

- On the supply side, fossil-based thermal plants become more efficient; an important share of the remaining capacity using fossil fuels is fitted with CCS, removing CO_2 from plants' emissions. Renewable supply (based on biomass, hydro, wind, sun, geothermal heat) accounts for 65% of electricity supply by 2050, against 20% today. With the contributions of nuclear and CCS-equipped plants, low-carbon technologies produce 94% of global electricity demand in this scenario (this is a global aggregate picture; national endowments and technology choices mean that not all options are available in all countries).
- On the demand side, electricity-using appliances and equipment become more efficient. Some of the other end-uses switch from fossil fuel to electricity use (e.g. transport, with plug-in hybrids or electric vehicles with a lower CO_2 content than conventional vehicles as electricity gets decarbonised).
- The annual electricity demand in the IEA's 2°C-compatible scenario is projected to grow at three times the rate of total demand over the period 2011-50: by 2050, electricity becomes the largest energy carrier, ahead of oil products, with more than 25% of the total final energy use, against 18% today (IEA, 2014a).

Figure 7.1. **Contributions to annual emissions reductions between a 6°C and a 2°C Scenario**

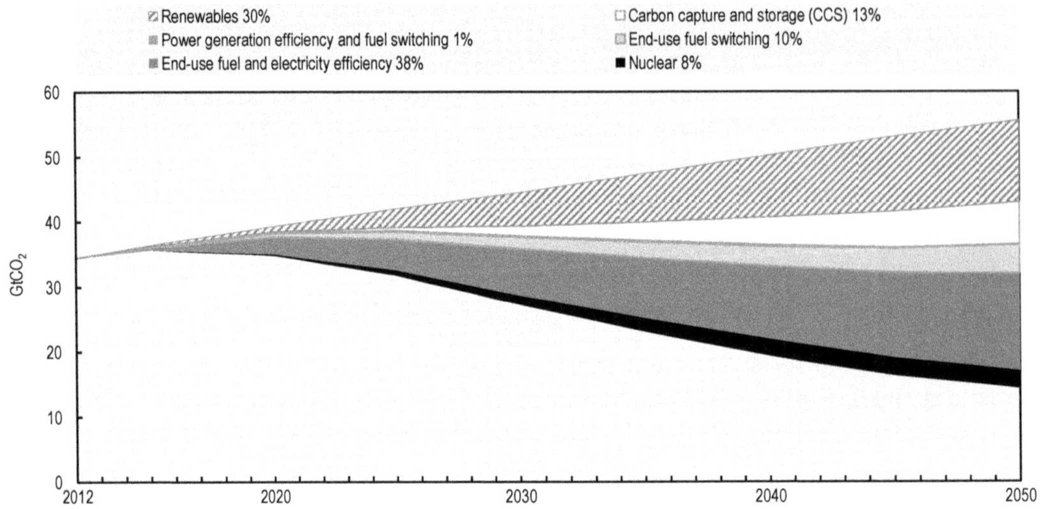

Source: IEA (2015a), *Energy Technology Perspectives 2015*, OECD Publishing, Paris, http://dx.doi.org/10.1787/energy_tech-2015-en.

Electricity therefore appears to be a driver of the low-carbon transition in the energy sector. Its expected role can be explained by: 1) electricity's credible aspiration to zero GHG emissions on the basis of existing technologies; 2) a strong government oversight of this activity; 3) the high value of electricity-related services to society, which may facilitate the acceptability of price increases. Other activities are also crucial for the low-carbon transition of the energy sector, such as mobility (see Chapter 8), industry and buildings. Heat is also an important energy carrier for the transition, including the potential it represents for energy efficiency improvements. Much of the energy used globally is still wasted in the form of heat, e.g. in thermal power generation but also in industrial uses; heat could also play a role in energy storage (IEA, 2014b). However, electricity presents an already visible strategic misalignment between regulatory frameworks in certain regions and their climate policy objectives.

Electricity is also an important source of human and economic progress through the many services it provides: lighting, refrigeration, water pumping, the operation of machinery, etc. as emphasised by the Sustainable Energy for All programme (SE4ALL, 2015). Giving access to this and other modern forms of energy is essential to bring people out of poverty and trigger development. Many solutions today are provided in a decentralised fashion (solar lamps, photovoltaic installations, etc.) but the urbanisation trend in developing countries hints at the growing role of centralised power systems in the future. Front-running regions and jurisdictions will set important examples if they can effectively trigger the required low-carbon generation investment.

How low-carbon technologies challenge the organisation of competition in electricity

The decarbonised electricity system in 2050 will depart from that of the 20th century. Recent trends and projections from the IEA and others indicate a growing share of variable renewable energy (VRE) i.e. based on solar and wind technologies, combined with so-called dispatchable, low-carbon sources (hydro-electricity, nuclear power, geothermal, biomass-fuelled thermal plants or storage) and, for some years to come, some fossil fuel capacity. While VRE technologies will not be alone in driving the decarbonisation of electricity systems, the fact that their supply varies and cannot be programmed puts some strain on electricity systems and markets, and demands in particular a more flexible and integrated system.[3] Another important feature of today's low-carbon generation technologies is their higher capital cost; these technologies, with the exception of biomass-fuelled and CCS plants, also have lower operational expenditures than fossil fuel generation technologies (standard coal, oil or gas plants).

Figure 7.2. **Electricity generation by technology and CO_2 intensity in the 450 Scenario, 1990-2035**

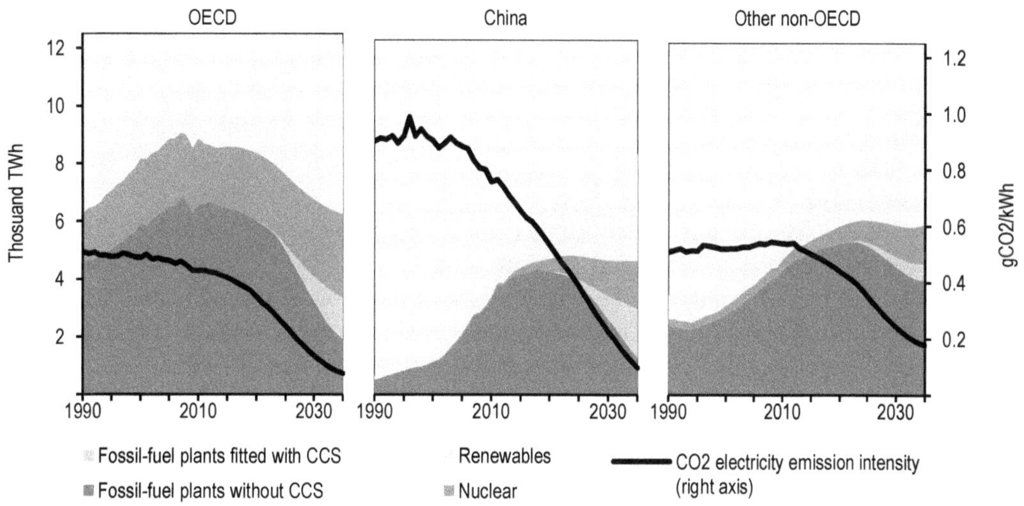

Note: CCS: carbon capture and storage.

Source: IEA (2014c), *World Energy Investment Outlook*, Special Report, OECD/IEA, Paris, available at: www.iea.org/publications/freepublications/publication/WEIO2014.pdf.

Today's markets and organisations of electricity systems may not be appropriate to drive an effective and least-cost transition toward a low-carbon electricity system. In particular, there may be a strategic conflict between the prevailing approach to electricity market liberalisation and CO_2 emissions reductions.

Under deregulated wholesale markets, the price is set by the variable cost of the marginal plant, i.e. the most expensive running plant mobilised to meet the full demand. With a market dominated by fossil fuel technologies with high variable costs and low fixed costs (such as gas turbines), new investments in baseload efficient plants benefited from higher electricity prices set by older, less efficient technologies. This has worked for new investment in baseload technologies, but to unlock future investment in plants that run fewer hours in the year, the system needs to experience periods with very high prices during scarcity hours.

However, a deregulated market left to its own devices will be unable to finance investments in low-carbon technologies with high fixed costs, which require long-term stability of electricity prices (see Keppler and Cometto, 2015). The price to pay would be very high carbon taxes and volatile prices with a substantial number of scarcity hours and price uncertainty, leading to higher-than-otherwise financing costs. Of course, core climate policies must step in to encourage low-carbon technologies, through market push (lowering technology costs) and pull (an increasing price on CO_2). In the medium to long term, the question is what kind of market-based electricity system could guide the decarbonisation of the electricity system on the basis of a CO_2 price signal, as opposed to *ad hoc* support measures in use today. The question is a useful starting point to identify misalignments in electricity system arrangements that could challenge the decarbonisation objective.

How is low-carbon electricity supported today?

> The ultimate target is to create conditions for market-based investments in low-carbon generation. (IEA, 2014a: 289)

In recent years, the decarbonisation agenda, combined with import dependency, diversity of supply and air quality concerns and innovation policy, have encouraged governments to support rapid deployment of low-carbon technologies. Wind plants and photovoltaic (PV) panels were the principal beneficiaries of out-of-market mechanisms such as feed-in-tariffs (FiT; a fixed price paid by rate-payers for the electric output of renewable installations, combined with priority access to the grid), or tax credits for investment in renewable energy capacity (i.e. a capital subsidy). Combined with a stagnant or declining electricity demand in Europe and the United States, the growth in renewable energy supply has led to the stranding of some conventional power plant capacity (IEA, 2014c).[4] Feed-in-tariffs have also been criticised on cost grounds and adjusted downward. This government-induced policy uncertainty has had detrimental impacts on investment (Baron, 2013). Governments have since introduced new instruments to better control costs and encourage a growing integration of variable renewables in electricity markets. On the upside, renewable support measures have triggered a noticeable change "from the classical electricity utility paradigm to a much less concentrated industry widely composed of highly speciali[s]ed firms" (Benatia, 2014).

The growing contribution of variable renewable energy requires a sufficiently flexible electricity system to accommodate these sources without jeopardising security of supply. In time, this means mechanisms providing for investment in flexible generation sources, demand-side management and response, and in grid management and interconnection, as well as in electricity storage, which could offer an important breakthrough.

Without out-of-market arrangements, which can include FiTs or long-term power purchase agreements (both of which may or may not include a subsidy element, i.e. a tariff higher than average costs), electricity markets today may only generate the following low-carbon investments: life-extension of nuclear or hydro-electric plants, and the conversion of a coal plant to biomass.[5] One can point to examples such as unsubsidised onshore and offshore wind farms and geothermal plants in New Zealand, but they correspond to exceptional resource conditions – including the absence of thermal fuel input in that country. Otherwise, investments in renewable, but also in nuclear capacity, are based on mechanisms that provide long-term income security in the form of a power purchase agreement, and not just through the operation of the wholesale market.

Mature market-based system vs. fast-growing "regulated" systems

The need for a rethinking of electricity market arrangements to facilitate a least-cost decarbonisation of electricity systems is elaborated in the following sections. The electricity market liberalisation initiated in the late 1980s has mostly occurred in some, but not all, OECD jurisdictions. These countries have also implemented very different policy instruments to engage the low-carbon transition of the electricity sector, ranging from the

European Union's Emission Trading System (EU ETS) combined with renewable energy to the regulation of GHG emissions from power plants in the United States.

In fact, much of the investment in electricity worldwide is undertaken by vertically integrated utilities or under regulatory arrangements that do not rely on wholesale electricity markets. For this analysis, a distinction is drawn between mature, market-based systems (most of the OECD) which sometimes face urgent investment needs to renew capacity, and fast-growing "regulated" systems whose primary challenge is to match a rapidly increasing demand for electricity.

How today's market structures risk hindering the decarbonisation of electricity

Diagnosing misalignment in regulatory arrangements

Market liberalisation – A brief history

While vertically integrated utilities still play an important role in some OECD countries, many adopted market approaches in the early 1990s, generally involving the unbundling of generation, transmission and distribution, and the opening up of competition, with the creation of wholesale electricity markets (sometimes called "energy-only" markets).

This competition benefited in part from the emergence of a new, flexible and more energy efficient technology: the combined cycle gas turbine (CCGT), with a low cost of capital and a smaller size than incumbents such as coal, nuclear power or hydro-electric plants.

Overall, liberalisation has improved the economic efficiency of the power generation, transmission and distribution sectors, even if competition is not as broad in the retail market as anticipated when reforms began. Wholesale electricity markets, however, have become critical in ensuring a flexible supply of electricity, in tandem with transmission system operators that manage the main electric trunks. These ensure that electricity is supplied at the lowest cost possible, as plants compete on the basis of their marginal cost of generation. Deregulated wholesale electricity markets are designed to optimise short-run dispatch. They allow maximising the use of existing assets and reducing overcapacity, which was one of the motivations behind liberalisation in the 1990s. They have, however, not been properly designed to ensure adequate investment in generation in order to ensure security of supply, especially in a context of uncertainties on climate policy. In the EU, the carbon price has not been enough to curtail investments in thermal fossil-fuelled power generation or to bring forward investment in low-carbon generation without government support of some sort.

The misalignment of wholesale electricity markets with decarbonisation

> Regarding the long-term decarbonisation target by 2050, Energy Technology Perspectives scenarios envision a share of zero or low marginal cost electricity of more than 80%. The current design of wholesale electricity markets might not provide the signals needed to trigger such investments. (IEA, 2014a)

Core climate policy instruments are critical to move power generation in the right direction. However, the cost of doing so may be significantly higher unless the underpinning market arrangements are aligned to reflect the nature of low-carbon technologies and their impact on electricity systems.

Wholesale electricity markets as we know them today no longer provide an adequate signal to investors in new capacity. This is generally referred to as the problem of "missing money", historically due to restrictions on electricity prices and well-known to electricity market policy analysts (see Finon, 2013; Baritaud, 2012):

- Pricing electricity on the basis of the variable cost of a marginal power plant does not guarantee the full recovery of capital costs for all power plants.
- Allowing prices to reach very high levels when capacity is scarce is a possible way to recover capital costs, but these episodes may be too few and uncertain to trigger investment in new capacity (IEA, 2014e).[6] By definition, during such episodes of scarcity, demand may not be fully met. In the absence of sufficient availability of flexible demand response, this will lead to politically and socially unpopular rolling brown-outs. At present, some governments seem keen to avoid such situations by preventing the closure of otherwise unprofitable plants (an arrangement referred to as "strategic reserves").

The missing money problem is greatly exacerbated by the profile of leading technologies for the decarbonisation of electricity: their higher capital-to-operations cost ratio than today's fossil-fuelled plants. Further, there are decreasing returns in the deployment of VRE plants, as they often operate at the same time (when the wind blows and the sun shines). Since they carry essentially zero marginal costs, VRE plants bring prices down precisely when they come into operation and could draw revenues from the market. This situation is sometimes described as wind turbines and solar-based technologies cannibalising their own market – the higher their deployment, the lower their market returns. With some exceptions, investment in low-carbon technologies largely hinges on out-of-market arrangements which frequently contain a subsidy element. The problem of price declines in the presence of VRE extends to other technologies which see their average revenue and load factors erode, leading to early retirements and temporary shutdowns (NEA, 2012). However, this would also occur with any technology supported outside the market.

Will a higher CO_2 price be enough to fix these problems? In principle, in the near to medium term it would increase electricity prices, as fossil-based plants will continue to be marginal suppliers on the market, granting a cost advantage to carbon-free sources. There is, however, a lack of visibility about electricity prices over the lifetime of a new plant.[7] This is an important barrier for investment in low-carbon, high capital cost technologies if it is to be paid back by wholesale electricity markets only. Again, the problem would also be exacerbated by the penetration of low-carbon technologies characterised by low or zero marginal costs, with depressing effects on market prices. Wholesale markets with the highest shares of variable renewable sources sometimes record zero or negative electricity prices when wind and solar power plants produce more electricity than the load.

The regulatory challenge for a decarbonised electricity system is therefore the organisation of competition in the future to ensure that the constraint on CO_2 emissions is reflected in both investment and operational choices of power generation, alongside other equally important constraints such as security of electricity supply. It is unlikely that such competition can be solely organised on the basis of variable-cost pricing and carbon pricing. The cost structure of low-carbon electric power technologies changes the requirements demanded of electricity systems, raising the question of how such structure should be reflected in the incentives provided by this sector.

Sketching the elements of a low-carbon electricity system

Pricing carbon

A price on CO_2 is necessary to ensure that operational decisions on the market reflect the cost of climate change, and discourage plants with a high CO_2 content from running. This is particularly important for dispatchable plants that will be required for baseload power generation when VRE plants do not generate power. A price on CO_2 will also give a cost advantage to these other low-carbon technologies, be they hydro-electric, nuclear, geothermal or biomass-fuelled thermal plants or fossil fuel plants fitted with CCS.

Sending the right investment signal

What market mechanism would be best suited to trigger investment in high capital cost and low operational cost technologies? A simple first answer could be to organise competition for the investment in new power plants, e.g. on the basis of the average cost of generated electricity, whereas today's investors take decisions based on the expected evolution of wholesale prices as set by the variable costs of different plants setting prices at different points in time.

An example of such an arrangement is the auction mechanism used in Brazil to procure new capacity in a specific location. Winners can be awarded a price for their electricity in a long-term contract, giving some certainty on expected returns during the lifetime of the investment. The Brazilian auction allows bidders to be compared according to their average cost of power: in the presence of a carbon price, a gas- or coal-fired power plant would bid with the CO_2 cost included, putting bidders using low-carbon technologies at an advantage. Bidders would also have to allow for fuel price risk, which could not be passed on to consumers in such an auction.

South Africa also adopted an auction mechanism for new capacity, including a contract for concentrated solar power with different prices paid depending on the time of supply, to reflect the cost difference between baseload and peak-time generation.

Some of the policy solutions to secure low-carbon capacity at least cost follow a similar logic. For instance, the European Commission's new state aid guidelines on renewable energy encourage reliance on tendering as an alternative to administratively set payment levels such as feed-in-tariffs, which have not always responded to rapid cost reductions and may not encourage the technical and financial innovation that can lead to cost reductions.[8]

Efforts are also made to integrate variable renewable generation in the electricity market by asking generators to put their electricity on the wholesale market – as opposed to granting priority access to the grid and guaranteeing out-of-market payment. For instance, feed-in premium systems demand that producers sell their power on the market and provide a payment on top. The UK Contract for Difference (CfD) is a variation on this arrangement: low-carbon technologies receive a variable premium on top of the received electricity price, up to an agreed price level. A CfD is also under elaboration to secure investment in new nuclear capacity. At this stage, however, CfDs are set on a negotiated basis and on a technology-by-technology basis. The first CfDs offered by the British government have been for nuclear power and offshore wind.

The deployment of large amounts of VRE such as wind or solar PV also requires that the right balance be struck between two priorities: a strong signal for investors, which can be based on average cost pricing, and an incentive for low-carbon generators to deliver their electricity when its value to the system is highest – as opposed to maximising total output. In the case of solar in regions relying heavily on air conditioning, this may include positioning panels to maximise production during times of peak demand. As individual households that consume and produce electricity from PV (so-called "prosumers") are not exposed to dynamic electricity prices, achieving this requires an adjustment in the support mechanism. One way of achieving the same result is to structure power purchase agreements with payments differentiated according to time of production, as in South Africa's auctions.

Securing flexibility and reflecting system costs

Effort is therefore still required to align investment signals with the need for fair competition across low-carbon and other electricity technologies. Low-carbon technologies play different roles in electricity systems. Hydropower, nuclear, biomass-fuelled and CCS-fitted plants are dispatchable – i.e. can generate when required. Variable sources such as wind and solar PV are not fully dispatchable. How easily they can be integrated depends on the degree of flexibility of the underlying electricity system and on the share and speed of deployment of VRE technologies (Box 7.1). If electricity demand is high in times without wind and sunlight, other capacity must be brought on. Incentives must be present for such capacity to remain available and be built in a timely fashion, whereas, as shown earlier, the penetration of VRE can lead to reduced revenues for these plants.

According to the IEA (2014d), if local concentrations ("hotspots") are avoided, VRE deployment has a negligible impact, at shares of around 2-3% of annual generation. Apart from small island systems, shares of 5-10% in annual generation will not lead to technical integration challenges if operations are adapted and VRE deployment patterns are well co-ordinated. For higher levels of penetration (25-40%), a system-wide transformation is necessary if excessive costs are to be avoided (ibid.). The magnitude of the system costs of a growing share of VRE is in debate, however (NEA, 2012; IEA, 2014d).

Jurisdictions have developed various instruments that could be used to complement variable resources with flexible plants. One widely discussed and increasingly implemented option is a capacity mechanism that secures revenues for plants to remain available. Other jurisdictions have adopted more targeted contracting to trigger new investments or relieve congestion in the system. So far, these mechanisms are used to ensure that generation can meet the peak of electricity demand, a legitimate issue for the low-carbon transition as carbon-intensive plants are being retired. The contributions of the demand side and storage to the integration of VRE plants are touched on below.

The integration of VRE plants also brings costs related to: 1) the short-term balancing of power, as variable plants come in and out; 2) the cost of backing up power: dispatchable capacity (i.e. power plants using fossil fuels, nuclear energy or dispatchable renewable-energy sources such as hydro-electric power, geothermal power or biomass-fuelled thermal plants) needs to be available when there is neither wind nor sun; however, this capacity is made less economic by the growing market share of variable plants; 3) the costs of connecting the best production sites to consumption centres and strengthening distribution grids to accommodate individual rooftop photovoltaic panels (NEA, 2012; IEA, 2014d).[9]

Given different resources and technology mixes envisioned in different countries, the cost of an electricity system transformation will vary widely, but it should be allocated fairly and transparently to avoid distortions.

Box 7.1. Adapting technical security of supply standards

The initial appearance of variable renewable energy in power systems necessitated the development of specific grid-connection requirements for these technologies to reflect their different capabilities and impacts on the electricity system. Early requirements were characterised by a "do-no-harm" approach, sometimes in the form of a simple disconnection of the variable resource, e.g. whenever frequency was too high. This initial response began to pose a problem for the system as the installed capacity of variable energy technologies increased significantly.

Under the grid code for German solar photovoltaic (PV) power plants, all plants were required to disconnect from the system if grid frequency rose above a level of 50.2 hertz, which may occur during a system disturbance. While such a rule allowed secure system operation at low penetration levels, as capacity increased, the disconnection of all solar PV at the same moment can put system security at risk. After this issue was identified, a retrofit programme was put in place to ensure that no sudden losses of generation through the disconnection of all solar PV capacity would occur as a result of grid code requirements.

Source: IEA, Renewable Energy Division, 2014.

Looking for misalignment beyond electricity generation

Transmission and distribution: How to mobilise least-cost potentials?

Transmission and distribution (T&D) networks and their regulation are a critical component of the security of electricity supply, including the deployment of VRE at least cost. Wherever variable energy resources will be called on significantly to reduce CO_2 emissions, T&D networks and their operations will need to adjust. A widespread transmission system is a clear asset to maximise the use of variable generation resources in broad geographical areas, and to ensure security of electricity supply in periods without wind and sunshine. Benatia et al. (2013) analysed the factors behind the productivity of wind plants based on a sample of 31 OECD countries. Beyond the obvious importance of each country's wind resource endowment, transmission capacity and cross-border trade are two of four elements of critical importance, and need to catch up with the deployment of new VRE and other low-carbon capacities (see also Bahar and Sauvage, 2013):

- the availability of dispatchable generation capacity, i.e. plants that can deliver electricity to the grid quickly to compensate for times when there is less wind
- electricity transmission capacity, i.e. the ability to move generated electricity over long distances inside a country
- energy storage, i.e. hydroelectric pumped storage has an increasing positive impact when wind capacity grows
- the possibility of cross-border electricity trade.

The IEA projects that USD 5.9 trillion must be invested in T&D between 2011 and 2035 in a 2°C-compatible scenario, of which 8% would be in infrastructure for the integration of variable renewables (IEA, 2014c). These investments are crucial for the success and cost effectiveness of the decarbonisation of electricity generation.

The policy alignment issue in the case of T&D stems from the more complex role of the grid in the low-carbon transition. A cost-effective response should reflect "the competitive interplay between network asset costs, network operational costs, electricity supply costs and other solutions such as demand response" (Volk, 2013).

In other words, a new planning framework is required to evaluate the cost and benefits of several technical solutions. A specific challenge for variable renewables is the proper allocation of the costs they impose on T&D. The unbundling of generation, transmission and distribution can sometimes complicate decision making in this area. An indicative planning, such as integrated resource planning in Australia and the United States can facilitate co-ordination.[10]

It is clearly important to reduce the policy costs associated with the move to low-carbon generation. Co-ordination of renewable and flexible resources over large geographic areas is one essential measure that can reduce the overall cost. The Agency for the Cooperation of Energy Regulators quantified the potential benefits associated with cross-border flows in Europe at several hundred million of euros per year (ACER/CEER, 2013). Power market integration, however, will require dedicated efforts, and strong policy commitments are needed to allow the development of more efficient markets over large geographic areas.

Electricity and heat storage

A potential breakthrough in electricity systems may come from the development of energy storage technologies for both electricity and heat services. The IEA's *Energy Storage* indicates the various technologies, uses and contribution of storage solutions to decarbonisation. The question is whether these storage solutions could seamlessly integrate today's electricity systems.

The current installed capacity of electricity storage, mostly in the form of pumped hydro-plants, is not new to electricity systems and amounts to 2.4% of the total installed capacity in power generation worldwide. In the specific European case, the pumped-hydro storage plants operated at night, absorbing excess power from nuclear plants and were remunerated through the sale of electricity at high prices during the day. This business model has been eroded by the drop in daily prices caused by overcapacity as demand stagnated and wind and solar power were deployed. There is in this case a gap between the private value of electricity storage – which decreased – and the public value of storage, which should increase as more variable resources are deployed. Policy makers must thus think about whether they want to provide added incentives to electricity storage for security-of-supply reasons.

Small-scale storage solutions are now commercially viable, e.g. in off-grid applications in combination with photovoltaic panels. Water heaters have been used in France, including through remote control, to both lower peak demand and store energy in off-peak times. This particular form of storage contributes significantly to lower peak demand in winter (5 gigawatts) which would otherwise be served by fossil fuel plants. New innovative storage solutions such as flywheels, the transformation of excess electricity into hydrogen and storage of heat in the form of molten salts in concentrated solar power plants are at various stages of development and commercialisation (IEA, 2014e).[11] Various storage solutions provide different services to the electricity system: some will store electricity across seasons, others from night to day, or for a few minutes or hours.

Storage solutions do not always mesh easily with pre-existing policy frameworks. For some time in the United States, transmission assets were forbidden to participate in wholesale electricity markets to avoid market manipulation. "This distinction between transmission and generation assets results in unintended negative consequences for energy storage technologies that can supply services in both the transmission and generation portions of the energy system." (IEA, 2014e: 12) The US Federal Energy Regulatory Commission has since made amendments to allow storage to receive compensation for supplied services. Also a lack of transparency on the cost of various electricity services prevents storage solutions from competing with generators.

Another key element for a cost-effective scaling-up of storage is time-of-use electricity pricing: storage can support the integration of variable resources by absorbing their supply when demand is relatively low (and so too electricity prices) and by selling the stored electricity at peak times. The possibility of using electric car batteries for this purpose is also being researched. The IEA (2014e) provides further recommendations on policy interventions to level the playing field for energy storage solutions, including heat. Indeed, much energy is still wasted in the form of heat, in industry or in thermal power plants, including nuclear plants.

The demand-side: In need of new policy frameworks

A more efficient use of energy is an essential element of the low-carbon transition, as discussed in Chapter 1. The global improvement in energy efficiency will come in response to: higher electricity prices; policies to address other market barriers to rational energy use, especially where price signals are imperfectly "received" by end-users; and structural changes away from more energy-intensive activities and behaviours.

These aspects will influence electricity systems over time. There are, however, opportunities for the demand side of electricity systems to become more responsive to the specific dynamics of a low-carbon energy system. Demand-side response (DSR, the active participation of end-users in electricity markets), demand-side management (DSM, the management of demand to lower energy supply investments) or energy efficiency could allow electricity systems to:

- Improve security-of-supply and decrease overall capacity by finding financial incentives to induce electricity users to either forego consumption (via demand reduction) or postpose consumption (via demand shifting) at times of peak demand.
- Avoid locking in new generation to respond to peak demand levels, traditionally supplied by natural gas or fuel oil-based turbines.

There is an actionable potential for both supply and demand in the residential and buildings sector, e.g. with space-heating, water heating and air conditioning. Battery storage in cars could also become a sizeable contribution of the demand side. Industrial actors ought to be well placed to respond to price incentives; it is less clear for households or small businesses, although there have been successful experiences with time-of-use pricing. Generally speaking, these electricity users are less exposed to changing electricity prices in real time or exposed to contradictory signals (Box 7.2). There are, however, technical solutions based on smart meters and automatic control devices that would go a long way towards increasing DSR in the residential and buildings sectors.

This particular area is not devoid of regulatory barriers, for example in the machine-to-machine communication essential for the deployment of smarter energy uses related to mobile-enabled applications (OECD, 2012).[12] Further, electricity retailers are sometimes reluctant to let consumers access the data from smart meters, preventing the connection to other systems that could facilitate electricity savings. There may be a regulatory gap to be closed in this area as well.

Box 7.2. When high electricity tariffs discouraged low-carbon alternatives

Standard electricity tariffs in Italy are progressive – the higher the level of consumption, the higher the price per kWh. This system was put in place to encourage an efficient use of electricity at a time when it was derived primarily from fossil fuels. Italy has achieved one of the lowest per capita consumption levels of electricity in the domestic sector.

In its recent energy strategy, Italy has identified home heating as a key area for decarbonisation of the economy and increased uptake of renewable energy – including electricity from wind and solar generation. Using heat pumps for space heating would be a way to use clean electricity for space heating while increasing overall energy efficiency. However, the progressive nature of electricity tariffs is an economic barrier to the penetration of heat pumps, as customers are charged a higher price of electricity upon installation. The Italian government has identified this problem and is now working to reform tariff structures.

Source: Italian regulator; IEA (forthcoming), *Italy 2015 Review: Energy Policies of IEA Countries*, OECD Publishing, Paris, forthcoming.

At the moment, DSR consists mostly of large industrial users that withdraw part of their demand against a market payment. Dynamic electricity pricing, a more direct instrument to encourage demand response, has been slow to develop in many countries. New business models have emerged, e.g. to aggregate and co-ordinate the response of several electricity users with the help of information and communication technologies. These companies work with industrial users to identify equipment that can adjust electricity demand at times of high electricity prices, leading to cost savings, and also provide other remunerated services to the electricity system. This market innovation is present in very few markets today and regulations often create barriers for its deployment (CEEM, 2014).[13] It is also important to ensure an efficient competition between DSR and generation capacity resources.

The IEA notes that legal and regulatory frameworks, retail markets and products are still not allowing full participation of the demand side, a clear case of policy misalignment (IEA, 2013b), all the more so as technical solutions now exist to exploit this potential. Furthermore, decentralised power generation (especially rooftop solar PV panels) is changing the status of some consumers into "prosumers" who may also want to maximise revenues from their generation equipment, including through better management of both their instantaneous end-uses and electricity output.

In summary: A regulatory transition underway?

Wholesale electricity markets born from the wave of electricity sector liberalisation will be increasingly challenged as the decarbonisation of electricity systems becomes an imperative. The high capital cost of low-carbon technologies, ranging from nuclear to VRE sources and CCS, requires a more stable price signal than that provided by marginal cost pricing. A price on CO_2, necessary in any case to discourage the operation of dispatchable fossil-fuel plants, cannot by itself deliver the robust signal for investments via wholesale markets. Investors will demand risk premiums that would add to the cost of capital and to the overall cost of the low-carbon transition. New arrangements are necessary to allow an efficient competition in investment on the basis of full generation costs – including the cost of CO_2 emissions and all other external costs.

As the new decarbonised electricity system will include a growing share of VRE installations, it must reflect the new costs these installations carry in terms of flexibility, grid strengthening, etc. to ensure an efficient allocation of resources. Transmission and distribution grids, the demand side, and in the longer run storage technologies, all face some policy barriers, i.e. misalignments hindering their contribution to the low-carbon transition. A systematic review of regulatory arrangements governing these activities and, when needed, regulatory reform will be important to facilitate the transition.

Misalignments in fast-growing and often regulated electricity systems

With higher population and gross domestic product (GDP) growth, emerging and developing countries will record much faster growth in electricity demand than OECD countries overall (Figure 7.2). Supplying affordable electricity to communities without access continues to be an international priority, as illustrated by Sustainable Energy for All and the proposed energy items in the Sustainable Development Goals (UN, 2015). To be consistent with lower global emissions in this century, countries outside the OECD will have to install the lion's share of low-carbon generation in coming decades. About two-thirds of capacity in nuclear, onshore wind and solar (photovoltaic and concentrated) power, and nine-tenths of coal plants fitted with CCS, would be built outside the OECD in 2050 in the IEA's 2°C-compatible scenario (IEA, 2014a).

The vast majority of electricity systems in fast-growing regions remain vertically integrated and do not rely on wholesale electricity markets as known in some OECD regions, even if some degree of electricity unbundling has occurred. In the People's Republic of China (hereafter "China"), the State Grid Corporation and the Southern Grids operate independently from power generation. There is also competition for new projects among several large generation companies, including companies that specialise in hydro-electric projects or nuclear. Most of the capacity remains state-owned. In India, about two-thirds of the generation capacity is government-owned (IEA, 2014c).

An appropriate alignment of incentives with the multiple goals of the electricity system will be critical if these regions are to achieve this transition at the lowest cost possible. What follows is a short, broad-brush description of how regulated and fast-growing electricity systems also face policy misalignments in spite of the centrally planned nature of their regulatory frameworks.

Fostering investment

Investors in low-carbon technologies (hydro, nuclear, new renewables) need an additional incentive when these technologies are not cost-competitive with fossil-based technologies. In many cases, subsidies are provided on a per-MWh basis, i.e. a payment for every unit of electricity delivered to the grid (equivalent to the feed-in-tariffs used in many OECD regions). In certain countries, investors can also access funding at relatively low interest rates, which constitutes an important cost advantage given the weight of capital expenditures in low-carbon technologies.[14]

There are, however, potential disadvantages for regulated fast-growing power systems specific to the low-carbon transition: the tendency to invest in excessive capacity, locking in fossil-fuel plants, and the difficulty for smaller companies to compete with vertically integrated utilities, whereas smaller projects based on wind or solar capacity could be borne by independent power producers and create a healthy competition on costs. These problems can be solved through specific legislation for independent power provision and a simple regulatory framework for new entrants (OECD, 2013b).

Investments in clean energy often take place in a situation of imperfect competition where a state-owned enterprise (SOE) is the incumbent. That context alone may not be a barrier to low-carbon investment, but there may be opportunities for enhanced cost competition. Governments can establish a level playing field between SOEs and private actors – including independent power producers and network operators – as well as between foreign and national actors in the electricity sector, in order to attract investment in low-carbon generation. In particular, policy makers need to ensure that:

- producers of low-carbon electricity benefit from non-discriminatory access to the grid, as uncertain grid access increases project risk
- investment in the grid is open to private investment
- private developers benefit from non-discriminatory access to finance, e.g. from state-owned banks
- tenders for public procurement are carefully designed with clear and transparent bid evaluation and selection criteria (OECD, 2013b).

Recent examples of successful tender auctions in countries with regulated markets, including Brazil, South Africa and the United Arab Emirates, show how well-designed competitive processes can be very effective in securing renewables generation capacity, and at record-beating low prices, in the mentioned examples.

Examples of misaligned incentives in operations

> While there is a policy in place to guarantee priority dispatch [of wind and solar power in China], grid companies do not always apply this in practice. (IEA, 2014d)

Where system operation and ownership of existing generation plants coincide there can be a significant conflict of interest: adding wind and solar PV to the system may compromise the economic value of existing assets. Permitting and grid-connection issues are likely to face lower barriers where generation, transmission and distribution have been unbundled. Furthermore, there is evidence that low-carbon resources may not always be used to their full capacity because not all stakeholders have incentives aligned with this objective. For instance, in China and India, the grid or utilities must purchase VRE at the premium price set by the government. As this electricity is more expensive than coal-based generation, a growing share of renewables implies a direct cost for the grid, as this cost is not automatically passed on to end-users through higher electricity tariffs. This creates a disincentive to accepting more VRE supply on the system.

The price that Chinese power generators receive for their output is determined administratively on a per MWh level. The general objective of such pricing is that generators can recover their average costs. However, this requires that generators produce the quantity of electricity planned for the year, such that total revenues match average costs. A similar approach is taken for the remuneration of land-based wind energy in China. The quantities of wind and coal electricity generation that have been procured in China with this remuneration scheme have not been fully aligned, however, and this creates competition. When wind generation is particularly high, the output of coal power plants is sometimes not reduced to the maximum extent possible, as this would reduce the sales of coal-based electricity and jeopardise the cost-recovery for coal generation. Wind generation is then curtailed, which affects its economics. This further results in an efficiency loss, given the much lower short-run costs of wind.

These are examples of misalignment between a policy goal (in this particular case, a growing share of wind power in overall supply) and the incentives generated by the existing regulatory structure. Utilities that are financially vulnerable will resist the integration of more expensive low-carbon sources of generation. Baron et al. (2012) have illustrated a similar conflict in the case of the now likely introduction of an emissions trading system in China's power generation sector, in which power plants generate on the basis of a negotiated annual output and receive a set price for the electricity that they put on the grid. The regulated structure of power generation will not deliver the efficiency gains expected from the CO_2 market mechanism unless some adjustment is made to the planning of power plants. In particular, it would be necessary to allow: 1) a pass-through of the CO_2 cost onto the price received by plants, to encourage competition from less carbon-intensive plants; 2) some flexibility for plants to adjust their running time to match their CO_2 emissions decisions.

Preliminary view on misalignments in fast-growing regulated systems

A much more systematic review would be necessary to identify whether vertically integrated electricity systems feature significant policy misalignments with the low-carbon transition, and determine whether more competition could facilitate the transition. The previous section indicated that not all forms of competition will necessarily send the right signal for investment in low-carbon, high capital-cost technologies. Further, if unbundling of generation, transmission and distribution delivers efficiency gains, it appears to make it more difficult to define an economically efficient strategy for the decarbonisation of power systems in which all options (generation, transmission, distribution, demand side) would contribute. This is clearly an area where jurisdictions with deregulated and regulated frameworks could usefully share experience.

Resilience of energy systems to climate change

An important exposure to gradual climate change and extreme weather events

Energy sector infrastructure stands to be affected by climate change, whether via extreme events, gradual temperature increases, sea-level rise or changes in precipitation, albeit with differences across regions (Figure 7.3). Recent extreme weather events have exposed the vulnerability of energy activities, but long-term projections indicate more wide-ranging effects:[15]

- Energy demand: changes in heating and cooling demand, especially in air conditioning, which relies overwhelmingly on electricity. This may lead to peak electricity demand in the summer; the IEA underlines that this could cause stress to the electricity system during heat waves. Increased irrigation needs caused by a warmer, drier climate will also drive up energy demand (IEA, 2013b).
- Oil and gas exploitation is often exposed to challenging weather conditions, and will need to adapt to gradual changes including: temperature increase (e.g. directly affecting the efficiency of natural gas liquefaction facilities); a sea-level rise; increased storm intensities; water stress (refineries and coal mines consuming large quantities of water); or permafrost thaw (leading to shifting pipelines and more limited use of ice roads). Climate change will therefore have a direct impact on infrastructure costs in this sector, not to mention possible losses of production. Although long ice-free summers in the Arctic open up exploration and shipping opportunities, "the technical and environmental challenges are already significant and a number of projects have either been held back by the complexity of operations and by environmental concerns, or suspended due to

escalating costs" (ibid.). Extreme weather events, as illustrated by Hurricane Katrina's damage to the oil and gas industry in the Gulf of Mexico, can also severely affect this sector, all the more so as 45% of recoverable conventional oil[16] is located in offshore fields.

- Electricity generation and networks are vulnerable to a warmer climate, but also to extreme cold weather (as experienced during so-called polar vortex incidents in North America in 2014 or the Canadian ice storm of 1998). Warmer air and water reduce the output of thermal power plants, whether they are fossil fuel-based or nuclear. Water cooling needs at power plants can be reduced, e.g. with dry cooling, but at additional cost. Electricity losses in transmission and distribution also increase with temperature, not to mention the vulnerability of these networks to extreme weather events. Recent research indicates that in the 2040s, a 1°C warming will reduce available electricity capacity during summers by up to 19% and 16% in Europe and North America respectively (ibid.).
- Turning to renewables, hydroelectricity will also be impacted, though sometimes positively (in Canada, Nordic countries and the Russian Federation). However, warming is more often associated with droughts, which drive massive reduction of hydropower generation in some locations. Wind patterns may change, affecting the productivity of existing farms; the occurrence of extreme wind and icing events is also a factor. PV-based electricity is also affected by increased air temperature and of course snow cover. Like other crops, biomass-based fuels will be affected by changes in yields driven by climate change, as well as by extreme weather events.

Can policy enhance the energy sector's resilience?

The energy "industry" is routinely confronted with weather-related incidents, which have increased in frequency and severity in recent years. Through experience, companies have developed contingency plans. They have been investing to either make their infrastructure more robust or ensure the ability of a facility to quickly come back on line after an incident (the choice between the two options being driven by cost considerations). At the same time, there is no certainty that the energy sector as a whole is taking into account the full measure of climate change impacts over the lifetime of its assets, or fully incorporating this dimension in investment choices.

The energy sector is explicitly addressed in some countries' adaptation strategies. For instance, policy measures are adopted to: 1) reduce climate risks (support for burying power lines, building sea walls, elevating or relocating important equipment like substations, establishing early warning systems, etc.); 2) transfer risks (e.g. public-private insurance); 3) absorb risks (a compensation scheme after an extreme event happens). An open question at this stage is whether there is a role for regulatory policy to steer energy sector stakeholders towards an enhanced resilience in the face of global warming impacts.

Figure 7.3. **Selected climate change impacts on energy**

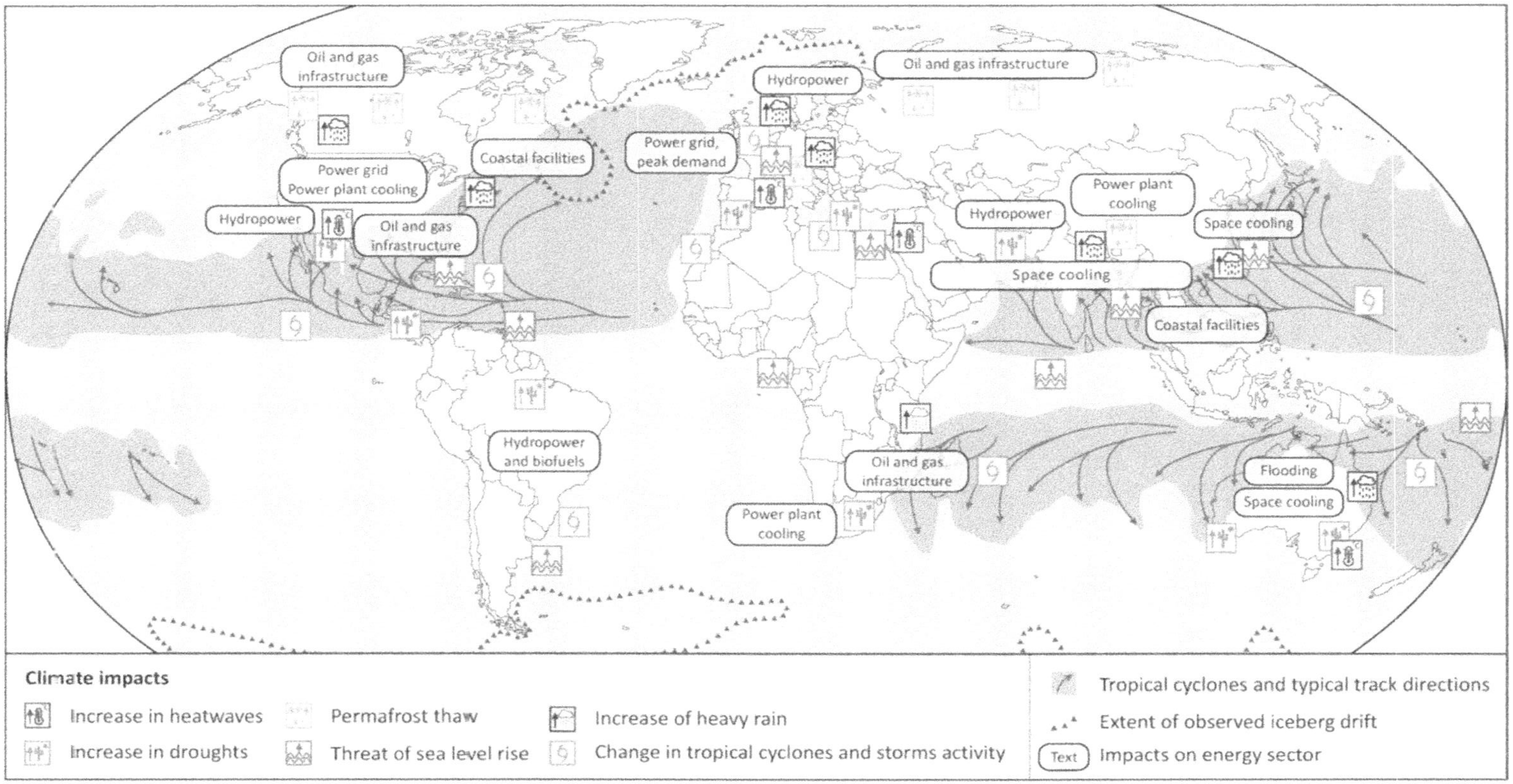

This map is without prejudice to the status of or sovereignty over any territory, to the delimitation of international frontiers and boundaries, and to the name of any territory, city or area.

Source: see IEA (2013b) for full references.

Zoning, insurance and water-use policies and technical standards[17] are sometimes used in this domain. The incentive to adapt effectively, however, hinges on other regulatory questions. For example, who bears the risk of supply disruption? What are the contractual arrangements in case of supply disruption? Do current forms of price regulation enable an efficient level of investment in resilience (e.g. burying transmission lines, ensuring sufficient redundancy)?

At this stage, some governments are engaged in discussions with the energy sector and related stakeholders (e.g. water utilities and users) to come to a better understanding of foreseeable impacts and response measures (see e.g. Zamuda, 2014). Canada's Adaptation Platform has engaged various public and private sector stakeholders to share and discuss climate and hydrology information, outcomes from risk assessments, best practices and tools, and policy drivers and barriers (Wilson, 2014).

There is mounting pressure on energy companies to report on the vulnerability of their assets to climate policy and climate change, although more with a view to attracting investors' attention to related risks than to fostering adaptation of those companies to climate change.[18]

In short, more work is needed to identify policy misalignments in the energy sector's resilience to current and future climatic change. The IEA is addressing this aspect through its Climate and Energy Security Nexus Forum (IEA, 2015b).

Notes

1. By 2050, according to the IEA's *Energy Technology Perspectives* 2DS Scenario, the power sector achieves similar levels of decarbonisation in the OECD, China and India, with reductions of 96-97% in CO_2 emissions from business-as-usual levels (IEA, 2014a).

2. These results are corroborated by the Deep Decarbonisation Pathway Project, which also emphasises the importance of decarbonising power generation and of electricity end-uses substituting fossil fuel uses to improve efficiency (SDSN and IDDRI, 2014).

3. With more decentralised sources of generation (PV rooftop panels), and the emergence of electricity-producing consumers ("prosumers"), the system also needs to change from a unidirectional to multi-directional mode. See also IEA (2013a).

4. In Europe, natural gas plants have stopped operating as the price of CO_2 allowances is too low to make gas plants competitive against coal plants. In Germany, coal plants also benefited from the closure of nuclear plants, with higher CO_2 emissions as a result.

5. The use of biomass in coal-based power plants has been supported in the EU, via the EU ETS, as have domestic support mechanisms for renewable energy.

6. The investment in gas and coal plants that took place in Europe in the last decade were motivated by then higher electricity prices and an underestimation of the share of renewable energy capacity going forward. The lower demand for electricity combined with a rapidly growing share of renewables have made baseload plants much less economical than anticipated.

7 This is illustrated by the drop in natural gas prices in the United States, or the recent drop in international oil prices. Such market surprises blur market expectations about electricity prices.

8. "Market instruments, such as auctioning or competitive bidding process open to all generators producing electricity from renewable energy sources competing on equal footing at the European Economic Area level, should normally ensure that subsidies are reduced to a minimum in view of their complete phasing out" (EC, 2014). The Netherlands has already introduced a feed-in-premium based on tendering (SDE+).

9. The shift to "prosumers" has a huge potential of mobilising goodwill and commitment at the grassroots level, but again poses the question of cost arrangements: should the cost of new investments needed in the distribution grid to include distributed generation be covered by prosumers or through generalised network tariffs? The answer could be important in determining the competitiveness of rooftop PV installations.

10. Resource means generation, demand response and networks.

11. A molten salts heat storage facility equips a 1 GW solar plant in Spain, for instance.

12. Other barriers of a technical nature stand in the way of DSR. One issue relates to machine-to-machine communication, which is an enabler for many "smart" systems such as smart grids and transport. A major barrier for machine-to-machine enabled mobile applications (and users) is the lack of competition once a mobile network provider has been chosen. The problem is the SIM card, which links the device to a mobile operator. By design, only the mobile network that sells an SIM card can designate which networks and devices a user can connect to. Changing SIM cards can be very costly; in some cases the user could be locked in a 10- to 30-year contract (OECD, 2012).

13. Examples include the Pennsylvania-Jersey-Maryland (PJM) system in the northeastern United States, and France and Norway.

14. It has been argued that wind turbines compete more easily with natural gas plants in Brazil's auction because they have access to low cost of capital via the country's development bank .

15. The IEA has also conducted several discussions among policy makers and the private sector on the Energy Security and Climate Nexus.

16. Excluding light tight oil.

17. The European Union has undertaken a mapping of technical standards related to infrastructure resilience (Paunescu, 2013).

18. In the United States, the Securities and Exchange Commission's interpretive guidance on climate disclosure seems to have had limited effects, with 59% of Standard & Poor's 500 companies reporting on climate. The quality of climate disclosure for those that do report is measured around 5 on a scale of 100 (Ceres, 2014).

References

ACER/CEER (2013), *Annual Report on the Results of Monitoring the Internal Electricity and Natural Gas Market in 2012*, Agency for Cooperation of Energy Regulators and Council of European Energy Regulators.

Bahar, H. and J. Sauvage (2013), "Cross-border trade in electricity and the development of renewables-based electric power: Lessons from Europe", *OECD Trade and Environment Working Papers*, No. 2013/02, OECD Publishing, Paris, http://dx.doi.org/10.1787/5k4869cdwnzr-en.

Baritaud, M. (2012), "Securing power during the transition: Generation investment and operation issues in electricity markets with low-carbon policies", *IEA Energy Papers*, No. 2012/13, OECD Publishing, Paris, http://dx.doi.org/10.1787/5k3wb8fmsz6h-en.

Baron, R. (2013), "Renewable energy: A route to decarbonisation in peril", paper prepared for the 29th Round Table on Sustainable Development, 4-5 June 2013, OECD, Paris, available at: www.oecd.org/sd-roundtable/papersandpublications/Background%20Paper%20RTSD%20June%202013.pdf.

Baron, R. et al. (2012), "Policy options for low carbon power generation in China: Designing an emissions trading system for China's electricity sector", *IEA Energy Papers*, No. 2012/12, OECD Publishing, Paris, http://dx.doi.org/10.1787/5k3wb8gl4l37-en.

Benatia, D. (2014), "The patterns of entry in the generation segment of the electricity supply sectors of OECD countries", paper for the Working Party on Integrating Environmental and Economic Policies, ENV/EPOC/WPIEEP(2014)8, OECD, Paris.

Benatia, D. et al. (2013), "Effectiveness of policies and strategies to increase the capacity utilisation of intermittent renewable power plants", *OECD Environment Working Papers*, No. 57, OECD Publishing, Paris, http://dx.doi.org/10.1787/5k46j0trlrnn-en.

CEEM (2014), " Compte-rendu du séminaire: L'Intégration de la 'demand-response' sur les marchés de l'électricité", Chaire European Electricity Markets, Fondation Paris-Dauphine, Paris, www.ceem-dauphine.org.

Ceres (2014), "Reducing system risks: The Securities and Exchange Commission and climate change", Ceres, Boston, Massachusetts, February, available at: www.ceres.org/files/investor-files/sec-guidance-fact-sheet.

EC (2014), *Guidelines on State Aid for Environmental Protection and Energy 2014-2020*, Communication from the Commission, European Union, available at: http://eur-lex.europa.eu/legal-content/EN/TXT/?uri=CELEX:52014XC0628%2801%29.

Finon, D. (2013), "The transition of the electricity system towards decarbonisation: The need for change in the market regime", *Climate Policy*, Vol. 12(2012), pp. 1-16.

IEA (forthcoming), *Italy 2015 Review*, *Energy Policies of IEA Countries*, OECD Publishing, Paris, www.iea.org/countries/membercountries/italy.

IEA (2015a), *Energy Technology Perspectives 2015*, OECD Publishing, Paris, http://dx.doi.org/10.1787/energy_tech-2015-en.

IEA (2015b), "Resilience of the energy sector to climate change", www.iea.org/topics/climatechange/subtopics/resilience (accessed on 23 March 2015).

IEA (2014a), *Energy Technology Perspectives 2014*, International Energy Agency, Paris, http://dx.doi.org/10.1787/energy_tech-2014-en.

IEA (2014b), *Linking Heat and Electricity Systems: Co-generation and District Heating and Cooling Solutions for a Clean Energy Future*, OECD/IEA, Paris, available at: www.iea.org/publications/freepublications/publication/LinkingHeatandElectricitySystems.pdf.

IEA (2014c), *World Energy Investment Outlook*, Special Report, OECD/IEA, Paris, available at: www.iea.org/publications/freepublications/publication/WEIO2014.pdf.

IEA (2014d), *The Power of Transformation: Wind, Sun and the Economics of Flexible Power Systems*, International Energy Agency, Paris, http://dx.doi.org/10.1787/9789264208032-en.

IEA (2014e), *Energy Storage*, IEA Technology Roadmaps, OECD Publishing, Paris, http://dx.doi.org/10.1787/9789264211872-en.

IEA (2013a), *Secure and Efficient Electricity Supply: During the Transition to Low Carbon Power Systems*, International Energy Agency, Paris, http://dx.doi.org/10.1787/9789264207639-en.

IEA (2013b), *Redrawing the Energy-Climate Map: World Energy Outlook Special Report*, International Energy Agency, Paris, available at: www.iea.org/publications/freepublications/publication/WEO_RedrawingEnergyClimateMap.pdf.

IEA (2012), *Electricity in a Climate-Constrained World: Data and Analyses*, OECD Publishing, Paris, http://dx.doi.org/10.1787/9789264175556-en.

IEA (2011), *Climate & Electricity Annual 2011: Data and Analyses*, OECD Publishing, Paris, http://dx.doi.org/10.1787/9789264111646-en.

Keppler, J.H. and M. Cometto (2015), "A review of nuclear new build with respect to project structure, supply chain and financing", presentation at the NEA Workshop on "Project and Logistics Management in Nuclear New Build", Paris, 11 March 2014, OECD, Paris, available at: www.oecd-nea.org/ndd/workshops/pmnnb/presentations/docs/presentation1.pdf.

NEA (2012), *Nuclear Energy and Renewables: System Effects in Low-carbon Electricity Systems,* OECD Publishing, Paris, http://dx.doi.org/10.1787/9789264188617-en.

OECD (2013a), *Effective Carbon Prices*, OECD Publishing, Paris, http://dx.doi.org/10.1787/9789264196964-en.

OECD (2013b), "OECD policy guidance for investment in clean energy infrastructure: Expanding access to clean energy for green growth and development", OECD Report to the G20, with contributions by the World Bank and UNDP, October, available at: www.oecd.org/daf/inv/investment-policy/CleanEnergyInfrastructure.pdf.

OECD (2012), "Machine-to-machine communications: Connecting billions of devices", *OECD Digital Economy Papers*, No. 192, OECD Publishing, Paris, http://dx.doi.org/10.1787/5k9gsh2gp043-en.

Paunescu, M.G. (2013), "The EU strategy on adaptation to climate change", presentation to the 3rd Forum of the Climate-Energy Security Nexus, 25 October, Paris, available at: www.iea.org/media/workshops/2013/nexus/Session5_3Paunescu_20131015_Presentation_IEA_slides.pdf.

SDSN and IDDRI (2014), "Pathways to deep decarbonization – Interim 2014 report", Sustainable Development Solutions Network and Institute for Sustainable Development and International Relations, available at: www.iddri.org/Publications/Pathways-to-deep-decarbonization-Interim-2014-Report.

SE4ALL (2015), Sustainable Energy for All website, www.se4all.org (accessed on 22 March 2015).

UN (2015), "Open Working Group proposal for Sustainable Development Goals", Sustainable Development Knowledge Platform, United Nations Department of Economic and Social Affairs, available at: https://sustainabledevelopment.un.org/sdgsproposal (accessed on 23 March 2015).

Volk, D. (2013), "Electricity networks: Infrastructure and operation. Too complex for a resource?", *International Energy Agency Insights Series 2013*, OECD/IEA, Paris, available at: www.iea.org/publications/insights/insightpublications/ElectricityNetworks2013_FINAL.pdf.

Wilson, A. (2014), "Canada's Adaptation Platform Experience – Enhancing energy sector's resilience to climate change", presentation at the 5th Nexus Forum on the Climate-Energy Security Nexus, November 2014, Natural Resources Canada, available at: www.iea.org/media/workshops/2014/5thnexusforum/4.3WILSONNRCan.pdf.

Zamuda, C. (2014), Presentation to the 4th Forum on the Climate-Energy Security Nexus: Energy & Water at the International Energy Agency and World Business Council on Sustainable Development, 12 June, United States Department of Energy, available at: www.iea.org/media/workshops/2014/nexus/P4CraigZamudaIEAWBCSD20140612.pdf.

Chapter 8

Opting for low-carbon urban mobility

Current transport systems rely largely on fossil fuels and impose very high environmental costs (climate change, noise, air pollution), particularly in urban settings. Policy intervention is needed to provide more energy efficient and less carbon-intensive mobility. These measures should focus on shifting away from the use of individual cars to mass transport modes, reducing the need for travel through land-use planning, as well as improving fuel and vehicle efficiency.

Sub-national governments are critical decision makers for low carbon mobility, but they face institutional, organisational and financial barriers to pursue ambitious climate action. These include co-ordination issues, contradictory incentives in the fiscal framework and a lack of capacity. Aligning policy action across levels of governments and between stakeholders could do much to deliver lower-carbon mobility. National frameworks and legislation should also give more financial and political leeway to local governments to pursue low-carbon choices.

Key messages

The reliance on fossil fuel-based transport systems has a very high local and global environmental price tag. The transport sector produces roughly 23% of global CO_2 emissions and is the fastest-growing source globally. Without further policy action, CO_2 emissions from transport could double by 2050.

Reducing emissions from transport could also mean cleaner air and less congestion, but this will not happen without proactive and joined-up policy action to:

- *avoid* unnecessary travel and reduce the demand for total motorised transport activity
- promote the *shift* to low-emission transport modes
- *improve* the carbon and energy efficiency of fuels and vehicle technologies.

These measures should be embedded in a strategy for urban development that makes more efficient use of space and takes into account environmental costs, well-being and economic development needs. This would mean refocusing urban transport policies on access rather than mobility itself, providing safe infrastructure for walking and cycling, shifting to mass transport modes where demand is concentrated and developing cities along mass transit corridors, improving fuel efficiency and promoting electric and fuel cell vehicles in concert with low-carbon electricity (or hydrogen) production.

Many cities in OECD countries are already designed around the private car. They will need very low – or even zero – emission fuels and vehicles, more efficient public transport systems, land-use planning that reduces the need for personal vehicles and alternatives to transport demand (such as teleworking) if they are to reduce CO_2 emissions significantly. In cities in developing and emerging economies, where much of the infrastructure is yet to be built, urban expansion needs to be managed in a way that limits the demand for energy-intensive mobility while promoting safe, affordable, accessible and sustainable transport systems for all. Sub-national governments are critical decision makers for urban transport planning, but co-ordination, general framework and capacity are generally barriers to sub-national governments making greater efforts toward climate action.

The roadmaps are reasonably clear, but several policy misalignments need to be corrected to allow urban mobility to develop while reducing its carbon footprint.

Are land-use planning and transport policies integrated at the metropolitan level? Land-use and transport planning are often the responsibility of separate authorities whose co-ordination mechanisms are limited or informal. This may result in development patterns that do not sufficiently account for transport needs, increasing inhabitants' reliance on personal vehicles. Mismatches between administrative boundaries and the functional extent of built-up areas also undermine joined-up planning. Solutions for better alignment range from the creation of metropolitan governance bodies that integrate land-use and transport planning to looser forms of inter-municipal and inter-sectoral collaboration such as contracts, platforms for dialogue and co-operation, and specific public investment partnerships.

Do cities have enough financial or political leeway to make low-carbon choices? A review of policy misalignments should involve a rigorous analysis of the impact of national policies on urban action. National legislation typically defines cities' and sub-national governments' responsibilities and revenue sources. The fiscal system dictates to a great extent what cities can and cannot do – as well as their incentives for action. How these revenue sources are designed can either enable or hinder sub-national governments' action towards low-carbon development. Some national tax provisions and regulations encourage further carbon-intensive development. Capacity building is also essential to help city governments design low-carbon cities, engage with the private sector and access funding.

Key messages (*cont.*)

Are policy signals aligned to facilitate the penetration of low-carbon breakthrough technologies? Our reliance on fossil fuel vehicles is perpetuated by the widespread use of fossil fuel-oriented infrastructure, innovation policies that support fossil fuels and the relatively low cost of road transport for consumers in some jurisdictions. These signals will need to be reversed for breakthrough low-carbon solutions to be rapidly deployed. Innovation in electric and hybrid vehicles is progressing, but a number of market failures and barriers are hindering the scaling-up of alternative fuel vehicles. Experiences with electric vehicles in the Netherlands and Norway show that combining battery-charging infrastructure, rebates on electric vehicle purchases and priority lanes on main access roads can lead to very rapid uptake. Such breakthroughs will require governments to provide the policy mix needed to challenge existing infrastructure based on fossil fuels.

The decarbonisation challenge for urban transport

Our modern and global economy has developed thanks to our ability to construct reliable and interconnected transport networks. For the vast majority of transport needs, this has entailed reliance on fossil fuels and the internal combustion engine. However, fossil fuel-based transport systems come with very high local and global environmental costs, and change is needed to achieve the low-carbon transition. If reducing emissions from aviation and shipping requires international co-operation (see Chapter 5), achieving low-carbon terrestrial transport systems will largely depend on how cities develop and how much more energy efficient transport systems will become. Sub-national governments are key actors in this low-carbon future, provided that their incentives are aligned with the climate challenge. A lack of co-ordination and of capacity, alongside other general framework conditions, creates barriers to sub-national governments making the most of public investment. These obstacles are particularly relevant for low-carbon mobility, which involves new technologies, skills and functions that often go beyond individual jurisdictions' prerogatives.

Transport is the fastest-growing source of CO_2 emissions globally

Greenhouse gas (GHG) emissions from transport represented 14% of global emissions in 2010 (IPCC, 2014), and 23% of CO_2 emissions from global fossil fuel combustion. Transport is the sector with the fastest-growing consumption of fossil fuels, and is also the fastest-growing source of CO_2 emissions globally (IEA, 2014). The number of individual cars increased from 625 million to 850 million between 2000 and 2013, mainly in developing countries (IEA, 2013). Aviation is also growing at a fast rate, though its share of global emissions is still limited to 2%. In 2012, road transport emissions represented more than 70% of CO_2 emissions from the entire transport sector. In the absence of policy action, CO_2 emissions from transport could double by 2050 (OECD, 2012c).

A radical and comprehensive approach to reducing emissions from transport is required (IEA, 2015) (Figure 8.1). This should simultaneously seek to:

- reduce the demand for total motorised transport activity (*avoid*)
- promote the use of "low-emission" transport modes such as walking, cycling and public transport (*shift*)
- use the most efficient fuel-vehicle technology system possible for all trips (*improve*).

Figure 8.1. **A potential contribution of strategies and transport modes to a 2°C future**

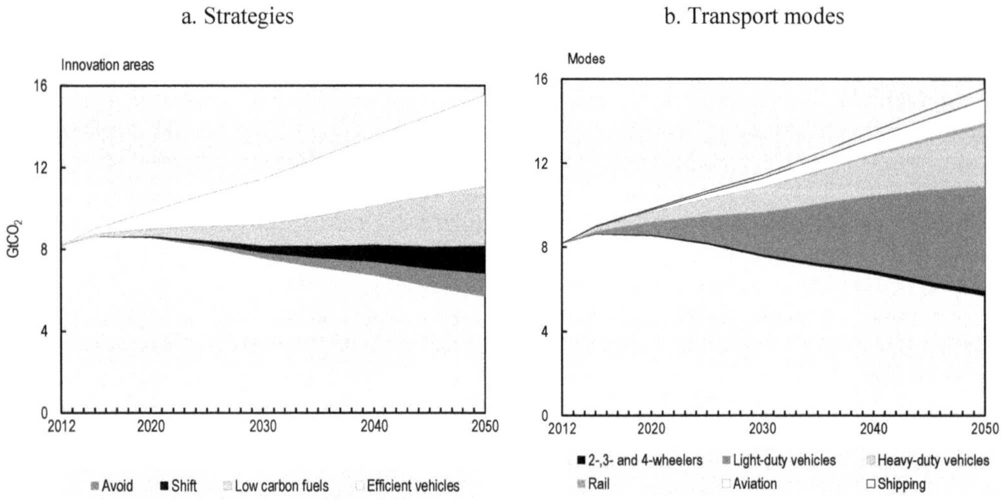

Source: IEA (2015), *Energy Technology Perspectives 2015*, OECD Publishing, Paris, http://dx.doi.org/10.1787/energy_tech-2015-en.

Low-carbon transport strategies should be embedded within a strategy for development that makes more efficient use of space and takes into account environmental costs, well-being and economic development. The challenge is particularly urgent in the context of the rapid pace of city growth in emerging economies, which is accompanied by high levels of air pollution. The latest estimates from the World Health Organization show that outdoor air pollution kills more than 3.5 million people per year globally, far more than previously estimated. The cost of air pollution in OECD countries plus the People's Republic of China (hereafter "China") and India is estimated at USD 3.5 trillion per year in premature deaths and ill health (OECD, 2014a).

Low-carbon mobility could mean cleaner air, less congestion and better growth. But this will not happen without proactive policy action and integrated approaches, particularly at the urban level.

Sustainable transport for low-carbon cities

Urban areas play an important role in current and future trends in transport emissions. By 2050, an estimated 66% of the global population is expected to live in urban areas (UN, 2014). Cities in the developing world are seeing the most rapid growth in car fleets, and it is here where population and income growth will be concentrated in the coming decades (Figure 8.2). For instance, 40% of the total global growth in CO_2 emissions from land-based passenger transport will be generated in urban agglomerations in Latin America, India and China (OECD/ITF, 2015a).

A city's low-carbon pathway will require a multi-dimensional approach, as it depends on several factors such as socio-economic variables, resource endowment, geographical settings, and financial and institutional capacities, and has specific challenges.

Figure 8.2. **Growth of world urban population in absolute numbers of new urban dwellers (billion inhabitants), 1950-2050**

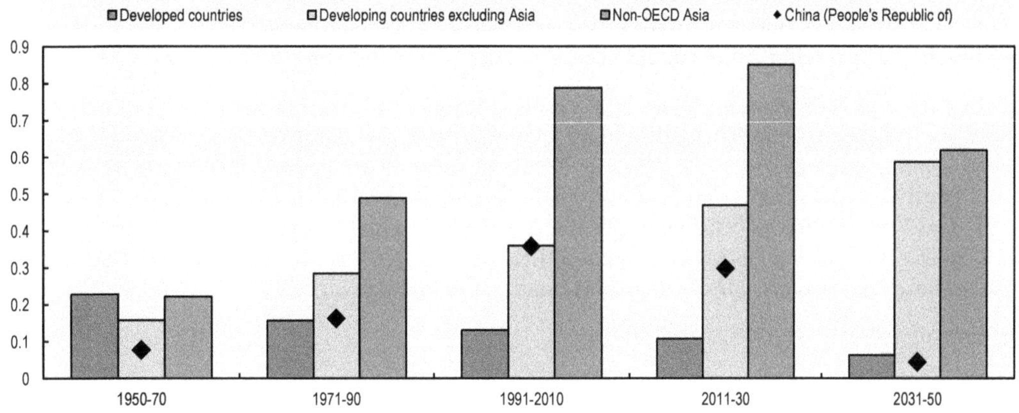

Source: UN (2014), *World Urbanization Prospects: The 2012 Revision*, United Nations, Department of Economic and Social Affairs, Population Division, Population Estimates and Projections Section, available at: http://esa.un.org/wpp.

For cities whose infrastructure is already locked in, very low – or even zero – emission fuels and vehicles will be key to a low-carbon strategy. This is the case in many OECD countries, where urban development has often involved low-density expansion, apart from some exceptions where it was geographically impossible (OECD, 2010). However, the need for integrated land-use planning is still a vital element of any policy mix aimed at minimising the predominance of automotive mobility in pre-existing urban settlements and reducing congestion. Complementary measures such as encouraging more efficient public transport systems, land-use planning or reducing transport demand through greater teleworking will further facilitate the low-carbon transition in these cities, while also contributing to reducing congestion and improving well-being (IEA, 2013).

In developing countries, where most of the infrastructure still needs to be built, there is more opportunity to influence the carbon signature of cities through land-use planning. Urban development needs to urgently limit the demand for mobility and be compatible with public transport while preserving well-being and economic growth and ensuring access to safe, affordable, accessible and sustainable transport systems for all, in line with sustainable development goals. “Avoid” and “shift” policies for passenger transport have the greatest potential in growing cities, where land-use planning and investment can still have an impact on the distance between homes and workplaces and the choice of transport on offer. Sprawl can be minimised and well-being maximised through carefully planned dense, mixed-use development with good mobility choices, while preserving green infrastructure. The International Transport Forum’s (ITF) analysis shows that promoting transit-oriented development could reduce the carbon intensity of urban mobility by 31% in Latin America, 26% in China and 37% in India, compared with a business-as-usual scenario (see OECD/ITF, 2015a; Box 8.1).

Policy makers will have to adapt their policy mixes to the physical characteristics and maturity of urban settings, as well as well-being and economic development factors. While compact city development can lead to fewer or shorter motorised trips and a greater use of public transport (OECD, 2012a), it can also have local negative impacts, such as increased congestion creating local concentrations of air pollution (Melia et al.,

Box 8.1. Promoting transit-oriented development would reduce the carbon intensity of cities

An ITF analysis shows that promoting transit-oriented development could reduce the carbon intensity of urban mobility by 26% in Latin America, 31% in China and 37% in India by 2050.

Transit-oriented development involves building mass transit corridors that serve as the main transport axes of cities while encouraging high-density development along them. As transit-oriented development increases the value of land along these corridors, it offers a better opportunity to finance investment through land-value capturing. Transit-oriented development also typically fosters mixed land use that tends to reduce the average distance of trips and facilitates non-motorised transport. By increasing the financial viability of public and non-motorised transport, as well as improving conditions for its use, transit-oriented development encourages the greater use of these less carbon-intensive transport modes in total urban mobility.

Two alternative policy pathways were modelled:

- A private transport-oriented urbanisation scenario was constructed involving policy trends that intensify the shift to private transport use. These are high sprawl, low expansion of public transport and low fuel prices. This scenario is combined with a scenario of rapid expansion of road infrastructure (high roads).
- A public transport-oriented urbanisation scenario involves policy trends that increase the role of public transport in urban mobility. This scenario combines low sprawl, high public transport expansion and high fuel prices. The policy pathway is modelled in a context of urban road infrastructure lagging behind urban population growth (low roads).

These pathways were compared to a baseline scenario constructed on the assumption that current trends in each of the contexts will continue in the future. Overall, aligning policies towards public transport-oriented urbanisation reduces the carbon intensity of urban mobility. This would cut transport-related CO_2 emissions growth by 31% in urban Latin America, 26% in China and 37% in India by 2050.

Figure 8.3. **Scenarios of growth in mobility and CO_2 emissions for Latin American, Chinese and Indian cities, in 2050**

2010=100

a. Growth in passenger kilometres

b. Growth in CO_2 emissions

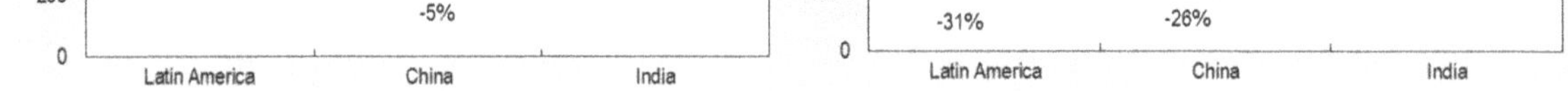

Transit-oriented development for new suburban areas must be carefully planned so that it does not create additional demand for development on land not served by the new public transport system and eventually lead to auto-dependent urban sprawl.

Sources: OECD/ITF (2015b), *ITF Transport Outlook 2015*, OECD Publishing, Paris, http://dx.doi.org/10.1787/9789282107782-en; OECD (2012a), *Compact City Policies: A Comparative Assessment*, OECD Green Growth Studies, OECD Publishing, Paris, http://dx.doi.org/10.1787/9789264167865-en.

2011), increasing heat island effects that can exacerbate climate change impacts, or destroying green urban infrastructure and ecosystems. Compact development can also face strong social opposition, as city dwellers may value green space and low-density housing. Where densification is deemed the way forward, it must be accompanied by measures to mitigate these local negative impacts. Such measures might include increased coverage by public transport systems or additional green spaces (OECD, 2012a).

Complementarities across policy sectors, programmes and levels of government

National governments have two clear roles to play in enabling sub-national governments to achieve their potential for advancing green growth: 1) to create favourable environments for sub-national action; 2) to integrate national and sub-national actions to improve coherence, promote learning, seek complementarities and reduce conflicts among sectoral strategies (GGBP, 2014).

Adopt a co-benefit approach to transport planning

Climate change concerns are rarely the catalyst of transport infrastructure decisions. These are usually driven by economic imperatives such as reducing congestion, improving accessibility and improving air quality in cities. Yet switching to low-carbon transport often has benefits for other policy goals. Such complementarities should be facilitated at higher levels of government by developing an integrated long-term infrastructure plan (Ang and Marchal, 2013).

Transport systems should also be adopted on the basis of cost-benefit analyses of alternative transport options and land uses which consider all the environmental, social and economic externalities of development. Economy-wide measures (e.g. carbon pricing) are critical to the response to climate change and for developing low-carbon transport. A strong national framework based on a carbon tax or price will broaden the range of environmentally effective options available to cities (OECD, 2014d). Many OECD countries have started using a price on carbon in their national cost-benefit analyses, albeit with very different values and methods (Smith and Braathen, 2015; see also Chapter 1).

Engage stakeholders and integrate transport and spatial planning to ensure effective provision of infrastructure

Whatever the city context, transport and spatial planning policies need to be co-ordinated across metropolitan areas to ensure that transport modes and land use work together to reduce GHG emissions. Co-ordinating land use and transport planning is also crucial to increase the climate resilience of new or existing transport infrastructure: land use and zoning can worsen or reduce its exposure and vulnerability to climate change impacts. Climate risk assessments and adaptation strategies need to be integrated into transport and land-use planning (Ang et Marchal, 2013).

Given the multiplicity of stakeholders and authorities involved in transport, aligning policy goals and integrating transport planning require systems of governance and institutional integration across levels of government and between stakeholders. Engaging key public, private and civil society stakeholders early in the decision making process of transport infrastructure investment is essential to ensure effective provision of sustainable transport infrastructure.

However, in the vast majority of cities, land use and transport planning are the responsibility of separate authorities whose co-ordination mechanisms are limited or informal. The result is suboptimal infrastructure design. In Bogotá, for example, limited planning for the optimal location of bus stations for the Transmilenio bus rapid transit (BRT) system and a lack of incentives for private property owners to redevelop land along corridors have hindered a low-carbon shift in urban development (OECD/ITF, 2015a). In order to reduce costs, bus stations were located in the medians of busy roadways, often with poor pedestrian access and in economically depressed and marginalised urban areas with minimal development potential. In Ahmedabad, land-use planning around bus stations was also absent in the planning process for the Janmarg BRT system. This has resulted in low pedestrian accessibility to BRT stops and limited land-use change around stations (Cervero and Dai, 2014).

The institutional structure and administrative boundaries of a metropolis often do not allow all aspects of public transport to be conceived and operated in a way that is designed to complement other modes and minimise overall environmental impact. Often functional areas – e.g. key commuting zones – do not match administrative units in urban areas (Figure 8.4). Such fragmentation can be a significant barrier to integrated transport strategies and promote increased car use. In Chicago, for example, approximately 36% of the population works outside the city, and 46% of city workers live in the suburbs. Yet the division of the public transport system into urban (Chicago Transit Authority, or CTA) and suburban (Pace and Metra) authorities means that CTA bus services typically end abruptly at the city limits where Pace services begin, and none of Metra's downtown commuter rail services connect directly to the CTA rail network (Merk, 2014). The result is greater car dependence. Fragmentation can be overcome through appropriate governance arrangements. In Paris, for instance, transport planning is developed at the regional level rather than at the city level, through the Syndicat des transports d'Île-de-France (STIF, Box 8.2).

Figure 8.4. **Functional areas do not match administrative units in urban areas**

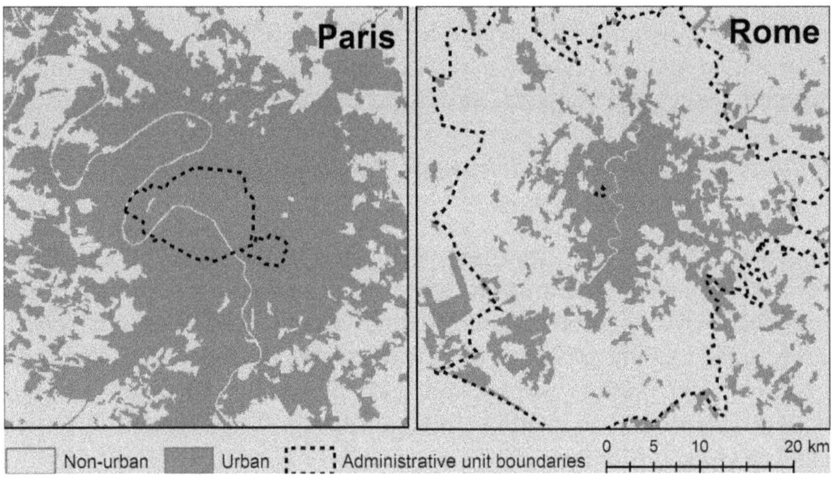

Note: The *OECD Metropolitan Database* identifies 1 148 urban areas, defined as integrated labour markets with at least 50 000 inhabitants in 28 OECD countries.

Source: OECD (2012b), *Redefining Urban: A New Way to Measure Metropolitan Areas*, OECD Publishing, Paris, http://dx.doi.org/10.1787/9789264174108-en.

Misalignment of pricing frameworks with low-carbon policy objectives also limits opportunities to cross-subsidise lower carbon-intensive modes with revenues collected from higher carbon-intensive modes. Cross-subsidising has proven a good way to increase the acceptability of transport demand management policies and increase funding availability. In other cases, fare structures are not integrated. In Jakarta, passengers often buy a new ticket every time they transfer BRT lines (Hidalgo, 2010). Weaknesses in the public transport system also reduce residents' mobility and limit the perimeter of their potential job markets, thus exacerbating inequality of access to employment within the metropolitan area.

Misalignment of objectives across different levels of government further increases complexity for integrated urban development planning. In Mexico, for example, municipalities are responsible for urban planning and development, whereas state governments are in charge of mobility and transport policy, transport infrastructure and operation of public transport. In the metropolitan area of Puebla-Tlaxcala, limited co-ordination between levels of government and a lack of integration between transport and urban development have contributed to the suboptimal siting of some BRT routes (OECD, 2013b).

Overcoming government fragmentation

Experience shows that there is no one-size-fits-all strategy for overcoming these administrative difficulties. Solutions for better alignment range from the creation of a single entity with authority over transport and land-use planning to looser forms of collaboration (OECD, 2010). Many urban agglomerations have created metropolitan governance bodies to integrate land-use and transport planning (Ahrend et al., 2014; OECD, 2015; see Box 8.2 for examples of successful transport and land-use integration). Other modes of co-ordination include contracts, platforms for dialogue and co-operation, specific public investment partnerships, joint authorities, or regional or municipal mergers (OECD, 2013c). Consolidating independent authorities on a metropolitan scale and building multi-modal planning and management capacity could help make significant progress towards an integrated transport and planning framework that fosters less carbon-intensive urban mobility.

To help resolve this functional mismatch, the OECD, in collaboration with the European Union, has constructed a new definition of cities that is comparable across countries and corresponds to their functional economic areas rather than administrative boundaries. The Functional Urban Area methodology uses information on density and commuting patterns in order to identify boundaries that approximate a city's functional labour market, relevant scale to plan transport and land-use planning (OECD, 2012b).

Remove barriers to action by cities

Sub-national governments have a fundamental role to play in the low-carbon transition. They are typically in charge of decisions that affect the carbon signature of cities, and can influence the total level of urban transport activity by changing the distribution of activities in space (for example, by changing land-use patterns, densities and urban design). In 2012, sub-national governments were in charge of 72% of total public investment (in terms of volume) across the OECD area. While not all of this is transport-related, 37% is allocated to economic affairs (transport, communications, economic development, energy, construction, etc.), around 23% is used for education, a further 11% is dedicated to housing and community amenities, and around 4% is dedicated to the environment (OECD, 2013f). All of these areas can be influential in transitioning to low-carbon cities. For example, education programmes can influence greener behaviours, and housing programmes can focus on energy efficiency.

Sub-national government budgets are under pressure following fiscal consolidation across all OECD countries. After a wave of "green" stimulus packages in 2008-09, resources for public investment have been reduced significantly, as they are one of the more flexible items in budgets. The challenge is not only to finance investments, but also to make the best of investment funds through appropriate governance arrangements. For instance, neither national nor sub-national governments may have had the appropriate tools or capacities to make the most of green stimulus funds.

The low-carbon transition requires that sub-national governments' financial incentives are aligned with the need to limit urban sprawl, encourage higher densities and provide public transport systems. As transport infrastructure investments are long-term, revenue needs to be predictable. Public transport infrastructure is also generally capital-intensive. Supportive national frameworks are vital for allowing cities to raise the revenue they need to create and maintain low-carbon infrastructure. These frameworks can also create the conditions – both in terms of legal competences and actual capacities – to enable cities to seek funding from the private sector and financial markets (OECD, 2010; 2014c).

Removing fiscal and regulatory impediments

National legislation typically defines cities' and sub-national governments' responsibilities, powers and, crucially, revenue sources. The fiscal system structures to a great extent what cities can and cannot do – and their incentives for action (OECD, 2014c). Despite local planning efforts to tackle urban expansion, some national tax provisions encourage carbon-intensive development (OECD, 2014d). These perverse incentives are discussed in more detail in Chapter 3, but the key steps needed to remove them follow.

- **Remove fossil fuel subsidies.** Fossil fuel subsidies favour and encourage the use of individual cars. In 2013, they amounted to USD 548 billion in developing countries (IEA, 2014). In OECD countries, between USD 55 and 90 billion was spent by OECD country governments annually between 2005 and 2010 to support coal, oil and gas production and use (OECD, 2013a).
- **Remove the tax differential between diesel and gasoline.** In 30 out of 34 OECD countries tax rates for diesel are lower than for petrol. This tax differential is at odds with the environmental effects of each fuel, including CO_2 emissions (Harding, 2014a).
- **Change the tax treatment of commuting expenses and company cars.** Company cars represent a substantial share of the car stock in many OECD countries. Tax regimes for company car use and commuting expenses can favour certain modes of transport over others, influencing how much, and how, employees travel (Harding, 2014b).
- **Fuel taxation: an incentive to sprawl?** Governments could investigate the extent to which fuel taxation has an influence on the type of urban forms. Fuel taxation is very different across countries, and can be an important driver behind the type of urban form (e.g. densely populated vs. sprawled metropolitan areas).
- **Reform property taxes to limit urban sprawl.** Property tax levied on buildings and other land improvements rather than on land values can make greenfield development more attractive to cities than infill, hence encouraging urban sprawl (Merk et al., 2012). In several countries, the design of property taxes promotes urban sprawl by favouring single-family, owner-occupied housing over multi-family or rental housing, or limiting incentives to move closer to work.

- **Remove perverse housing and zoning regulations.** Individual cities' housing and zoning policies (including restrictions on density, floor-to-area ratios and the construction of multi-family housing) can strongly influence GHG emissions (Glaeser and Ward, 2009). Limits on building height in cities and zoning that blocks denser development in the suburbs can raise housing prices. This can cause workers to seek housing further away from their jobs where it is more affordable. This then tends to increase carbon emissions from transport unless significant investments in public transport are made. In many countries, e.g. Chile or Mexico, national urban housing authorities also commonly provide low-income housing on city peripheries, where land is cheaper but access to public transport is poor and financially unsustainable (OECD, 2013c).

Ensure a resilient and stable tax base

For cities in the OECD, taxes are the most important source of revenue, representing more than half of all revenues on average. Sub-national governments need a broad and diverse tax base if they are to be resilient to economic fluctuations. Local revenues most often come from highly volatile taxes, such as corporate profits. Corporate tax revenues are mobile, highly cyclical, geographically concentrated and prone to shift the tax burden onto non-residents (Blöchliger and Petzold, 2009). Property taxes are more stable than corporate taxes, and their design can be instrumental for limiting sprawl. For example, split-rate property taxes – in place in Sydney; Hong Kong, China; and cities in Denmark and Finland – tax land more heavily than buildings (see also Chapter 3). Alternatively, some countries have transfer mechanisms that incorporate automatic stabilisers, such as those used in northern Europe to finance social services (Council of Europe, 2011).

Box 8.2. **Examples of integrated land-use planning and transport at the metropolitan level**

United Kingdom: Transport for London

Transport for London (TfL) is a statutory body created by the Greater London Authority (GLA) Act 1999. This act gives the Mayor of London a general duty to develop and apply policies to promote and encourage safe, integrated, efficient and economic transport facilities and services to, from and within the Greater London area. TfL is in charge of the operation of public transport, highway construction and management including congestion charging and vehicle licensing. In order to align an increasing population with more sustainable low-carbon transport, TfL investments range from introducing cleaner buses to encouraging alternatives to motorised travel such as cycling and walking, as well as improvements to the urban environment. The financial capacity for continuous investment and improvements comes from six main sources: income from fares and the congestion charging scheme; central government funding; a proportion of London business rates; prudential borrowing; commercial development of its estate, including advertising and property rental; and development and third-party funding for specific projects (TfL, 2014; ITF, 2013).

Box 8.2. **Examples of integrated land-use planning and transport at the metropolitan level** (*cont.*)

United States: Metropolitan Planning Organisation

In the United States, urban areas with more than 50 000 residents must have a designated metropolitan planning organisation (MPO) in order to qualify for federal transport funding. To obtain federal funding, MPOs are required to produce long-term transport plans based on a comprehensive analysis of demographic, travel and employment trends for their regions, as well as to propose a series of transport improvements to meet projected needs. Each decision on major investments planned for the region must be evaluated against a set of alternatives in order to ensure that the most cost-effective solutions are chosen. The long-term plans are then translated into rolling five-year transport improvement programmes, which list all projects to be funded in the MPO's jurisdiction over the next five years and identify the sources of funding that have been allocated to each. This certain (long-term) funding stream relieves the financial strain on metropolitan authorities in their quest to meet increasing needs for metropolitan-wide mobility and encourages co-ordinated decisions (OECD, 2015).

Singapore: Land Transport Authority

In Singapore, the Land Transport Authority (LTA) is a statuary board under the Ministry of Transport, established in 1995 as a result of a merger of four government agencies. A higher degree of integration was achieved by removing administrative boundaries. The LTA is responsible for planning, operating and maintaining land transport infrastructure and systems, including road safety, vehicle licensing and electronic road pricing. In order to reduce carbon emissions, the LTA constantly improves and expands its current public transport network, complementing it with parking policies and electronic road pricing. This has seen a shift from private to public transport use over recent years. In 2012, 63% of trips during morning and evening peak times were on public transport, compared to 59% in 2008. To increase this number even further, the LTA aims to make the transport system even more accessible and competitive. By 2030, 80% of households should be within a 10-minute walk of a train station, 85% of public transport journeys (less than 20 kilometres) will be completed within 60 minutes and 75% of all journeys in peak hours will be via public transport (LTA, 2013). The LTA's investments mainly stem from government grants and operating income, such as management fees from the government (taxes, fees and charges relating to land transport services) (LTA, 2014).

Germany: Rhein-Main-Verkerhsverbund

In Germany, every large urban agglomeration is covered by a transport authority. These transport authorities usually bring together all sub-national governments located in the metropolitan area as well as the relevant federal government departments. They manage public transport provision across the various modes of transport, provide strategic planning and co-ordinate pricing schemes for tickets that are valid across different modes of transport and different service providers. Typically, the transport authorities cover the full extent of the metropolitan area, but in some cases they reach significantly beyond their borders. In the Frankfurt metropolitan area, the Rhein-Main Transport Association (*Rhein-Main Verkehrsverbund*, RMV), brings together 3 levels of government: 15 counties, 11 cities and the federal state of Hesse. All key decisions on policy and strategy are taken by the Supervisory Board of the RMV, which agrees on the range of services on offer and the fare system. Its members are delegates of the shareholders, with each having one vote in order to adequately represent the different ideas and needs of the individual regions in the RMV network area. To achieve low-carbon emissions, the RMV also invests in cleaner trains and buses; in 2011 hybrid buses were introduced to the transport network. The RMV is financed by fare revenues, federal funding transferred via the state governments and municipal funds (OECD, 2015; RMV, 2014).

France: Syndicat des transports d'Île-de-France

Organising transport authorities (known as AOT in French) have existed in France since 1980 with the aim of promoting urban transport alternatives to private cars. The STIF is the organising transport authority for the Île-de-France and is integrated by the Region of Île-de-France, "departments" (counties) within the region and the city of Paris (CEPAL, 2009).

Box 8.2. **Examples of integrated land-use planning and transport at the metropolitan level** (*cont.*)

The institutional setting of the STIF is similar to other metropolitan areas of France, but it has had more success than most other transport authorities in achieving institutional co-ordination to advance low-carbon mobility goals. For example, between 2001 and 2010 total public transport trips in the Île-de-France region rose 21%, while car trips only grew by 0.6% (STIF, 2012).

The STIF "defines general operational and service level targets, sets fares and negotiates performance-based contracts with public service providers". The STIF is responsible for negotiating contracts with the national government (*contrat de project État-region*, or CPER) in which they agree upon a programme for capital investment and set funding responsibilities between national and regional governments. The STIF also develops an urban mobility plan (*plan de déplacements urbains*, PDU) that includes land-use and transport plans that guide all subordinate levels of government. The PDU includes precise objectives that should be met in order to contribute to the national CO_2 emission mitigation target (20% reduction by 2020) and establishes a programme of actions that will help attain such objectives.

This programme of actions is subject to opinions from regional, general and municipal councils; transport users; experts and environmental associations (the STIF approves the plan after one month of public consultation). This framework has been important in fostering consensus over future plans among all actors and has also facilitated public support. It has been important in giving coherence to the multiplicity of projects implemented by different authorities (since only those that are coherent with the plan can be financed) (CEPAL, 2009).

Curitiba, Brazil: Urban Transport Planning Authority

The Curitiba Metropolitan area comprises 26 municipalities with a total population of over 3.2 million. Curitiba's bus rapid transit system (BRT) is renowned as a pragmatic, integrated, cost-effective and efficient transport system. The city's success is thanks to close co-operation between its urban transport planning authority (URBS), responsible for transport, and the urban development authority (IPUCC), which is in charge of land-use planning. Since its launch in 1974, the Curitiba BRT has been notable for its careful alignment with the 1965 Curitiba Master Plan, which focused the city's growth along major corridors through land-use and zoning regulations. Building on this core concept, the Curitiba BRT has continually improved efficiency and expanded access. The integrated transport network (RIT) brings together feeder and inter-district buses, with transfer stations and a single fare, and has considerably improved the system's coverage and utility. Today, the RIT covers 14 of the 26 cities which make up the metropolitan area (URBS, 2014). The RIT was conceived around structural axles that provide the backbone of a transit-oriented development initiative through relatively low-cost and high-impact interventions (ITDP, 2014). Despite Curitiba's above-average rate of car ownership, the BRT service combined with parking policies has reduced automobile trips per year and ambient air pollution is among the lowest in Brazil (Lindau et al., 2010). The URBS is responsible for the planning, management, operation and control of the transport system in Curitiba. It defines routes, capacity and schedules; regulates and controls the bus system; and collects all fares. Bus operations are contracted to private sector operators. Fare revenues are pooled and paid to the contracted operators on the basis of the services provided. The complete RIT system, with its range of buses and integrated flat passenger fare, is reported to operate without subsidy (Lindau et al., 2010).

Source: OECD/ITF (2015a), "Shifting towards low carbon mobility systems", Discussion Paper No 2015-17, May 2015 OECD/ITF, Paris.

Increase cities' autonomy

Transfers from national governments are another significant source of revenue for sub-national governments, but they have pros and cons. Low-carbon transport investments would typically be eligible for earmarked grants and matching, as they generate positive spillovers for neighbouring areas (OECD, 2014b). Government transfers are also useful when risk sharing is needed, or when guidance from higher levels of governments is needed to improve co-operation and align different policy objectives (OECD, 2009), and when it is desirable to align priorities across levels of government.

However, reliance on transfers from higher levels of government often limits cities' scope for medium- to long-term planning means they relinquish part of their fiscal autonomy, and renders them vulnerable in times of economic difficulty (OECD, 2014c).

Problems can also arise from a lack of co-ordination among funding from a variety of national funds or between these and local funds. In Mexico, for example, 13 funds are available to finance urban mobility, but there is no co-ordination among them. Federal grants to local projects do not include sustainable urban mobility objectives, which means investment in urban roads is prioritised. In 2013, 74% of federal funds for urban mobility in 59 metropolitan areas were spent on road infrastructure (ITDP, 2014). Conditioning the allocation of grants to environmental objectives could facilitate the shift to a low-carbon economy. For instance, some *Länder* (states) in Germany have included ecological functions in their calculations of the fiscal transfer from *Länder* to sub-national governments (OECD, 2010).

Since the financial crisis, many cities around the world have had to cut budgets following a fall in intergovernmental transfers (Merk et al., 2012; OECD, 2013f). Between 2007 and 2012 in the OECD area – especially Iceland, Ireland, Italy, Portugal and Spain – sub-national governments' direct per capita investment contracted sharply (-7% in real terms between 2007 and 2012 and -15% in the three years since). In most countries, sub-national governments decided to preserve current expenditure on welfare, health or education and cut public investment (OECD, 2014c).

Reform fees and charges to discourage sprawl and encourage public transport use

Fees and charges are a third source of revenue for sub-national governments. For example, development charges can be levied on project developers to help fund the underlying infrastructure. The design of these charges can also encourage certain types of development (such as brownfield projects in city centres to avoid urban sprawl; OECD, 2010). While in some municipalities development fees increase with distance from the urban core, more often their rates are similar across the city, which does nothing to discourage sprawl.

Transport-related fees and charges can play an important role in reducing car use, particularly congestion charges, parking fees and taxes. Congestion pricing and other forms of road user charges (e.g. highway tolls) can help reduce congestion, vehicle use, local air pollution and GHG emissions. Such concepts have been implemented in London, Stockholm and Singapore. Congestion pricing is usually handled locally, imposed for entry to downtown and business districts, and sometimes calculated based on the congestion level or time of the day (Ang and Marchal, 2013). Road pricing systems (e.g. fixed road network access charges, tolls and electronic kilometre charges) can also target trucks to encourage freight efficiency improvements and shift from trucks to rail. Such policy should be part of an integrated strategy that proposes alternatives to the use of individual cars such as mass transport systems and promotes teleworking practices. Carbon and congestion pricing schemes are typically set to internalise external costs such as GHG emissions or travel time costs resulting from congestion. As such, those tools are designed to be demand management systems rather than revenue-generating instruments. However, revenues are often earmarked: parking charges are also frequently used as sources of revenues by sub-national governments, e.g. to finance public transport (Ang and Marchal, 2013).

Land sales are a large source of revenue for certain cities, particularly in India and China (Box 8.3). For example, land sales represented 80% of local revenues for the city of Shenzhen in the 1990s (OECD, 2013e; Merk et al., 2012). However, these policies can encourage cities to convert arable land into urban land with a view to selling it, thereby encouraging low-density urban expansion. A more environmentally friendly urban planning framework would require this type of funding arrangement to be reformed.

Box 8.3. How sub-national governments' fiscal frameworks influence urban development in China

In China, municipal finance depends heavily on land-related income such as land leases, auctions and development. This has created strong financial incentives for sub-national governments to build excessive quantities of scattered, low-density urban development.

Moreover, cities have a financial incentive to make a large amount of land available for industrial development in order to attract investment. This restricts land availability for residential and service sectors and contributes to unchecked urban sprawl, high housing density coupled with an oversupply of land available for industrial development. A tradition of functional separation in zoning, in turn, leads to longer commutes and more congestion.

Sources: OECD (2013c), *Green Growth in Cities*, OECD Green Growth Studies, OECD Publishing, Paris, http://dx.doi.org/10.1787/9789264195325-en; OECD (2013f), *OECD Regions at a Glance 2013*, OECD Publishing, Paris, http://dx.doi.org/10.1787/reg_glance-2013-en; Merk, O. et al. (2012), "Financing green urban infrastructure", *OECD Regional Development Working Papers*, No. 2012/10, OECD Publishing, Paris, http://dx.doi.org/10.1787/5k92p0c6j6r0-en.

Facilitate informed access to financial markets

The new context of fiscal austerity means that it is fundamental for cities to be able to access capital markets to fund low-carbon development. When funding is limited, investment in transport systems, in particular public transport facilities, is often the first to suffer. Several large urban areas throughout Europe and North America have relatively mature public transport networks whose operation and upkeep is expensive. The Chicago region, home to the largest rail and bus networks in the United States, illustrates the downward spiral that occurs when budgets run low: its public transport infrastructure has suffered such major underinvestment in maintenance that existing and future services are compromised. Financial sustainability has also been elusive, and the current system – based largely on fare and sales tax revenue – will struggle to keep up with rising expenses (ITF, 2013).

To overcome this shortage of funding, sub-national governments need to be able to leverage private sector investment and access international financial markets. However, the ability of cities to raise financing on international markets is limited. Sometimes national frameworks do not allow sub-national governments to issue bonds, and undermine local financial stability by restricting sub-national governments' ability to raise taxes and borrow from the private sector. In addition, sub-national governments often have lower credit ratings than the national government, as their default risk is considered to be higher (Merk et al., 2012; see Box 8.4).

Box 8.4. Improving cities' access to credit and finance

Access to credit and finance is a key challenge for cities in financing their infrastructure base. The nature of the challenge varies with country conditions, and the instruments and tools used to facilitate access to finance will differ also. In low-income developing countries, grants, loans and other development finance instruments could be relevant.

One of the key problems for developing country cities is their poor creditworthiness. The World Bank estimates that only a small percentage of the 500 largest cities in developing countries are deemed creditworthy – about 4% on international financial markets and 20% on local markets. The World Bank and its partners have designed a City Creditworthiness Program to help city financial officers conduct thorough reviews of their municipal revenue management systems and take the first steps to qualify for a credit rating. To address immediate financing needs, cities, with the support of national governments or international financing institutions, can access new or innovative financing arrangements such as loans, bonds, specific investment funds, tax arrangements or market-based mechanisms that may be particularly useful to finance green investments. For example, the US federal government financially supports municipal bonds (through tax exemptions and subsidies), and the World Bank offers green bonds for cities in low- and middle-income countries as part of the project financing component within a country's assistance portfolio. Sub-national governments should use innovative financing instruments with an understanding of the capacities needed, as in some cases they could compromise local finances and cause risky dependence on financial markets.

Sources: Based on Word Bank City Creditworthiness Program; Merk, O. et al. (2012), "Financing green urban infrastructure", *OECD Regional Development Working Papers*, No. 2012/10, OECD Publishing, Paris, http://dx.doi.org/10.1787/5k92p0c6j6r0-en; OECD (2013g, *Government at a Glance 2013*, OECD Publishing, Paris, http://dx.doi.org/10.1787/gov_glance-2013-en.

Between 2007 and 2012, sub-national gross debt per capita in the OECD area grew by 14%, corresponding to an increase of around USD 1 000 per capita. Only Israel, Switzerland and the United States saw sub-national government debt decrease on average during that period (OECD, 2013g). When monitoring municipal debt, national governments face a trade-off. On the one hand, cities need room for manoeuvre, especially in times of financial stress, to invest in infrastructure. On the other hand, many countries have run into difficulties over sub-national government debt.

The main challenge, therefore, is to create mechanisms that ensure economic stability and sound fiscal management while ensuring sub-national governments have sufficient financial capacity to deliver public services and finance essential investment (OECD, 2014d). So that national governments can ease the way for city governments to develop and implement their own financing strategy (OECD, 2014b), clear policies on prudential and macroeconomic limits to borrowing will be needed and subjected to processes of public scrutiny. Clear rules and procedures are also needed to govern cases of municipal insolvency (Council for Europe, 2011).

Build capacity for working with the private sector

Involving the private sector in urban transport infrastructure can contribute to the success of the low-carbon transition. For example, many mass rapid transit systems require close collaboration between public and private actors. While public-private partnerships (PPPs) and other frameworks for private participation in the supply of local public goods and services are increasingly important in many OECD cities, experience to date with PPPs has been mixed. While they should deliver better value for money if properly designed, they have sometimes been used to finance projects that would not have been approved otherwise. An OECD survey found that the lack of adequate capacity by sub-national governments for managing PPPs was seen as a significant challenge by 16 of 19 responding OECD country governments (OECD, 2013d).

Governments wishing to expand the role of private finance in urban public investment may therefore need to address these capacity issues. They could do so, for example, by establishing dedicated PPP units to work with local authorities, establish guidelines for lower level governments or create explicit incentives to engage the private sector, such as minimum private sector funding to access national funds (OECD, 2014c).

Removing bottlenecks to energy efficiency and technology breakthrough

Our reliance on fossil fuel vehicles is perpetuated by the widespread use of fossil fuel-oriented infrastructure, innovation policies that support fossil fuels and the relatively low cost of road transport for consumers. These signals will need to be reversed for breakthrough low-carbon solutions to be rapidly deployed. Innovation in electric and hybrid vehicles is progressing, but a number of market failures and barriers are hindering the scaling-up of alternative fuel vehicles. Such breakthroughs will require governments to provide the policy mix needed to challenge existing infrastructure based on fossil fuels. Experiences with electric vehicles in the Netherlands and Norway show that combining battery-charging infrastructure, rebates on electric vehicle purchases and priority lanes on main access roads can lead to very rapid uptake. However, these efforts come at a high cost. There is a need for a thorough cost-benefit assessment of those programmes. Policy flexibility is also advised in order to avoid technological "lock-in".

Energy efficiency could be the most important contribution to low-carbon transport sector

Energy efficiency from technology improvements could be the most important contribution to GHG emission reductions in transport by 2050 (IEA, 2015). A strong and stable price on carbon would be the first best solution to move towards more efficient transport modes, but many other barriers stand in the way of energy efficiency. Policies to improve road vehicle fuel economy should address market failures, information gaps and the higher upfront costs associated with more innovative technologies.

An integrated approach to improving the efficiency of new cars combines three policy elements: information measures such as fuel economy or CO_2 emissions labelling; vehicle fuel economy and CO_2 emission standards; and fiscal measures such as vehicle taxes and tax incentives and fuel taxes (IEA, 2012).

Many OECD and non-OECD countries have started developing policy support for fuel efficiency, but their policies are not always in line with the climate objective. The appropriate mix of policies depends on a particular country's circumstances, such as the overarching policy setting, infrastructure, and market and behavioural failures that need to be addressed (IEA, 2012).

Fuel efficiency standards are useful tools for delivering fuel efficiency and generating the technological changes needed for a low-carbon transition, provided that they are technology neutral. Fuel economy standards have been implemented in most OECD countries, as well as in China and India. But countries should investigate whether these standards are in line with long-term climate goals. Because of an expedient policy process and strong vested interest, targets may be set at the most feasible level rather than the optimal level necessary to meet the objective. Standards can also sometimes be designed to favour national car industry. In addition, some gaps remain for heavy duty road vehicles. Note should be taken of the multiplication of different standards internationally – this could create fragmented markets for the private sector, limiting incentives to investment.

Aligning vehicle taxes with the fuel efficiency or CO_2 emissions values of vehicles can make fuel-efficient vehicles more fiscally attractive, strongly encourage consumers to buy such vehicles and incentivise manufacturers to improve their fuel efficiency. While in the past vehicle taxes were mostly based on engine capacity, power or vehicle mass, in the last few years many countries have switched their vehicle tax system to account for CO_2 emissions. Some countries combine vehicle taxes with a rebate to customers purchasing lower fuel-consuming vehicles, such as the *bonus-malus* system in France (IEA, 2012).

Overcoming infrastructure lock-in: the case of alternative fuel vehicles

It is impossible to predict the technology mix that will achieve the low-carbon transition. But it is clear that while existing technologies offer potential for further improvements in environmental performance, new low-carbon solutions and more radical innovations will have to enter the mass market (Beltramello, 2012). This will require governments to remove support to existing carbon-intensive technologies and bring economic, legal and regulatory frameworks in line with the demands of new technologies and business models.

Governments should consider implementing a range of policy interventions, depending on the maturity of the technology considered (IEA, 2013a: Box 2). For high-potential and lower maturity technologies, technology push policies could be justified in a context where urgent climate action is needed: such policies include research, development and demonstration (RD&D) support and support to the deployment and development of energy distribution infrastructure, particularly in urban contexts. They would also lead to reduced fuel tax revenues, making the political economy of implementation challenging.

Innovation in electric and hybrid vehicles has been progressing more rapidly than other green innovation (Figure 8.4), but a number of market failures and barriers hinder the scaling-up of these alternative fuel vehicles. These include the lock-in created by the existence of infrastructure network effects to fuel different vehicle types; consumer reluctance to adopt "innovative" vehicles whose characteristics have not been fully demonstrated in the market; and limited financing for high-risk investments (such as those associated with R&D in alternative fuel vehicles; Haščič and Johnstone, 2012).

The Netherlands and Norway have very actively promoted electric cars and have achieved the highest penetration of electric cars in the EU. With already more than 31 000 electric vehicles as of 2014, the Netherlands' programme to promote electric cars was well ahead of its objective of 15 000-20 000 electric vehicles on the road by the end of 2015. In Norway, 14% of all new cars registered were electric in 2014. Incentives to facilitate the take-off of electric cars include: financial incentives to equalise the price between electric cars and conventional vehicles, information initiatives to reduce potential perception barriers and increase social acceptance, and other benefits such as access to bus lanes, free road tolls or ferry charges, and free public parking (OECD/ITF, 2015c). Still, these efforts come at a high cost and the effectiveness in terms of GHG emission reductions is relatively low. More work is needed to evaluate the overall effectiveness in terms of cost-effective climate solutions of such policies.

Figure 8.5. **Inventive activity in technologies included in the "green" patent index, 1990-2010**

World total, change on 1990, high-value inventions PF2

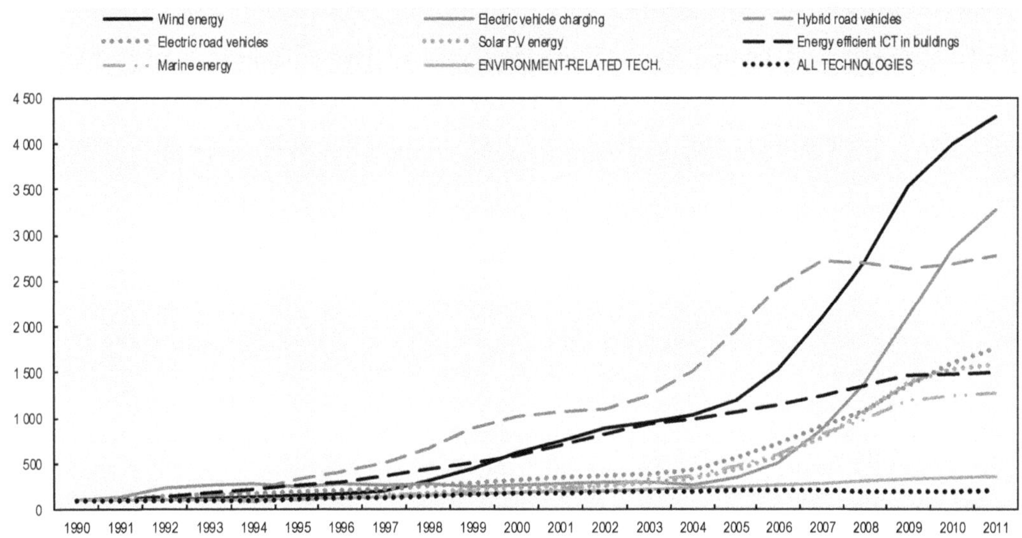

Note: Only patents for which protection has been sought in at least two international patent offices (PF2) are included. This has been found to be a good proxy measure for high-value patents.

Source: Haščič, I. and M. Migotto (2015), "Measuring environmental innovation using patent data: Policy relevance", *OECD Environment Working Papers*, OECD Publishing, Paris.

To overcome these barriers, government policies need to provide incentives along the spectrum from invention to commercialisation and diffusion. Policy support is critical, especially to ensure that vehicles become cost-competitive, to provide adequate recharging infrastructure and to influence consumer behaviours to change vehicle purchase and travel behaviour. A multi-pronged approach is required, targeting both the supply side (technology push) and the demand side (market pull), but this could be costly.

Policies to support technological development:

- Tailor government support policies to different stages of technology development and base them on assessments of expected costs and expected benefits – taking any interactions with other instruments into account (OECD, 2012c).
- Ensure environmental standards are sufficiently stringent to "force" technological change (Box 8.5).
- Design policies that provide incentives for innovators to drive emissions down to zero. Such policies have the potential to encourage "radical" and hitherto unforeseen innovations.
- More research, development and deployment is needed to reduce battery costs for electric vehicles, as well as into "smart" grids (IEA, 2009).
- Avoid using policy to "pick winners" (e.g through public procurement, R&D support standards) and instead ensure policy flexibility to keep open a wide spectrum of technological options, including for alternative fuel vehicles relative to other types. This will avoid early technological lock-in. It may be more efficient to support general infrastructure or technologies which benefit a wide range of applications, such as improved energy storage and grid management in the electricity sector.

- Ensure that design of incentives allows for a competitive selection processes, focuses on performance, avoids vested interests and includes evaluation of policies (Haščič and Johnstone, 2012).
- Maintain continuous commitment to the policy objective: to take the risks required, innovators need a credible and predictable policy framework.
- Public procurement could play an important role in scaling-up sales in electric vehicles. It would help overcome barriers such as network effects, economies of scale, demonstration effects and consumption externalities (Erdlenbruch and Johnstone, 2003).

Box 8.5. A mix of policies is required to force technological change

Haščič and Johnstone (2012) analysed the relative impact of prices and standards on environmental innovation in motor vehicles. The analysis found that while changes in relative fuel prices encouraged innovation in hybrid propulsion, performance standards were necessary to drive the development of technologies related to purely electric vehicles. For example, in order to induce a 1% increase in electric vehicle innovations, the stringency of California's Zero Emission Vehicle (ZEV) standard would have to be increased by 2%, while to induce a 1% increase in hybrid vehicle innovations, fuel prices would have to increase by 5% (Figure 8.6).

Figure 8.6. **The effect of prices and standards on alternative fuel vehicle innovation**

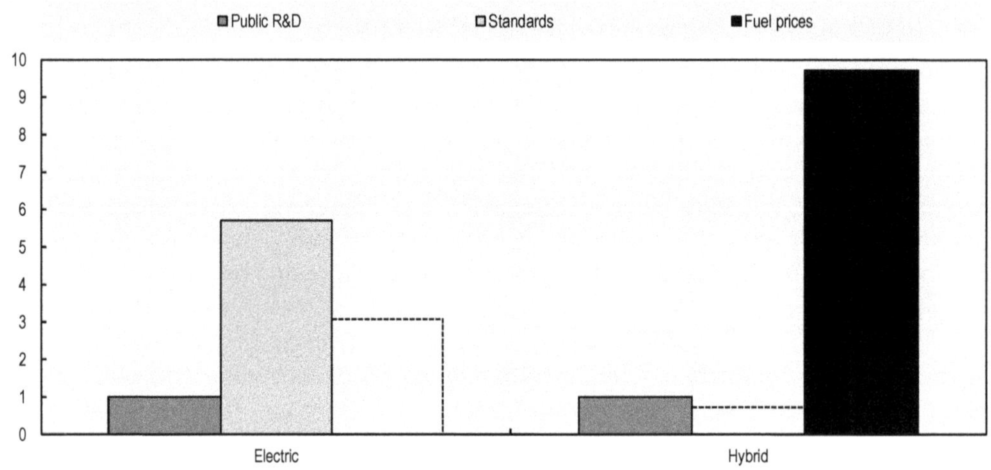

Note: The histogram shows empirical elasticities, evaluated at sample means and normalised in terms of the effect of "public R&D spending" (R&D = 1.0). Bars shown "without fill" represent estimates that are not statistically significant at the 5% level.

Source: Haščič, I. and N. Johnstone (2012), "Innovation in electric and hybrid vehicle technologies: The role of prices, standards and R&D", in: OECD, *Invention and Transfer of Environmental Technologies*, OECD Publishing, Paris, http://dx.doi.org/10.1787/9789264115620-5-en.

References

Ahrend, R. et al. (2014), "The OECD Metropolitan Governance Survey: A quantitative description of governance structures in large urban agglomerations", *OECD Regional Development Working Papers*, No. 2014/04, OECD Publishing, Paris, http://dx.doi.org/10.1787/5jz43zldh08p-en.

Ang, G. and V. Marchal (2013), "Mobilising private investment in sustainable transport: The case of land-based passenger transport infrastructure", *OECD Environment Working Papers*, No. 56, OECD Publishing, Paris, http://dx.doi.org/10.1787/5k46hjm8jpmv-en.

Beltramello, A. (2012), "Market development for green cars", *OECD Green Growth Papers*, No. 2012/03, OECD Publishing, Paris, http://dx.doi.org/10.1787/5k95xtcmxltc-en.

Blöchliger, H. and O. Petzold (2009), "Taxes and grants: On the revenue mix of sub-central governments", *OECD Working Papers on Fiscal Federalism*, No. 7, OECD Publishing, Paris, http://dx.doi.org/10.1787/5k97b11972bn-en.

CEPAL (2009). Políticas de movilidad urbana e infraestructura urbana de transport, United Nations Publishing. Cervero, R. and D. Dai (2014), "BRT TOD: Leveraging transit oriented development with bus rapid transit investments", *Transport Policy*, Vol. 36, pp. 127-138, November, http://dx.doi.org/10.1016/j.tranpol.2014.08.001.

Council of Europe (2011), *Local Governments in Critical Times: Policies for Crisis, Recovery and a Sustainable Future*, Council of Europe, Strasbourg, France.

Erdlenbruch, K. and N. Johnstone (2003), "Introduction", in: OECD, *The Environmental Performance of Public Procurement: Issues of Policy Coherence*, OECD Publishing, Paris, http://dx.doi.org/10.1787/9789264101562-2-en.

GGBP (2014), "Green growth in practice: Lessons from country experiences", Green Growth Best Practice, Seoul, available at: www.ggbp.org/sites/all/themes/ggbp/uploads/Green-Growth-in-Practice-062014-Full.pdf.

Glaeser, E. and B. Ward (2009), "The causes and consequences of land use regulation: Evidence from Greater Boston", *Journal of Urban Economics*, Vol. 65(2009), pp. 265-278, http://dx.doi.org/10.1016/j.jue.2008.06.003.

Harding, M. (2014a), "The diesel differential: Differences in the tax treatment of gasoline and diesel for road use", *OECD Taxation Working Papers*, No. 21, OECD Publishing, Paris, http://dx.doi.org/10.1787/5jz14cd7hk6b-en.

Harding, M. (2014b), "Personal tax treatment of company cars and commuting expenses: Estimating the fiscal and environmental costs", *OECD Taxation Working Papers*, No. 20, OECD Publishing, Paris, http://dx.doi.org/10.1787/5jz14cg1s7vl-en.

Haščič, I. and N. Johnstone (2012), "Innovation in electric and hybrid vehicle technologies: The role of prices, standards and R&D", in: OECD, *Invention and Transfer of Environmental Technologies*, OECD Publishing, Paris, http://dx.doi.org/10.1787/9789264115620-5-en.

Haščič, I. and M. Migotto (2015) "Measuring Environmental Innovation Using Patent Data: Policy Relevance", OECD Environment Working Papers, OECD Publishing (forthcoming). Hidalgo, D. (2010), *Modernizing Public Transportation: Lessons Learned from Major Bus Improvements in Latin America and Asia*, World Resources Institute, Washington, DC, available at: www.wri.org/sites/default/files/pdf/modernizing_public_transportation.pdf.

IEA (2015), *Energy Technology Perspectives 2015*, OECD Publishing, Paris, http://dx.doi.org/10.1787/energy_tech-2015-en.

IEA (2014), *World Energy Investment Outlook*, Special Report, OECD/IEA Publishing, Paris, available at: www.iea.org/publications/freepublications/publication/WEIO2014.pdf.

IEA (2013), *A Tale of Renewed Cities: A Policy Guide on How to Transform Cities by Improving Energy Efficiency in Urban Transport Systems*, Policy Pathway, OECD/IEA, Paris, available at: www.iea.org/publications/freepublications/publication/Renewed_Cities_WEB.pdf?dm_i=UP4,1PR31,7TKVE4,63KV2,1.

IEA (2012), *Fuel Economy of Road Vehicles*, IEA Technology Roadmaps, OECD Publishing, Paris, http://dx.doi.org/10.1787/9789264185029-en.

IEA (2009), *Electric and Plug-in Hybrid Electric Vehicles*, IEA Technology Roadmaps, OECD Publishing, Paris, http://dx.doi.org/10.1787/9789264088177-en.

IPCC (2014), Intergovernmental Panel on Climate Change, Working Group III, Summary for Policymakers, www.ipcc.chITDP (2014), *Invertir para Movernos: Diagnóstico de Inversiones en Movilidad en las Zonas Metropolitanas de México 2011-2013*, Mexico.

ITF (2013), "Funding urban public transport. Case study compendium", International Transport Forum, Paris, available at: www.internationaltransportforum.org/pub/pdf/13Compendium.pdf.

Lindau, Luis Antonio, Dario Hidalgo, and Daniela Facchini (2010) "Curitiba, the cradle of bus rapid transit." Built Environment 36.3: 274-282. LTA (2014) Annual Report 2013/2014, www.lta.gov.sg/content/dam/ltaweb/corp/PublicationsResearch/files/AnnualReports/1314/LTA%2 0AR%202014%20web.pdf (accessed November 2014).

LTA (2013) Master Plan 2013, www.lta.gov.sg/content/dam/ltaweb/corp/PublicationsResearch/files/ReportNewsletter/LTMP2013Report.pdf (accessed November 2014).

Melia, S. et al. (2011), "The paradox of intensification", *Transport Policy*, Vol. 18, No. 1, pp. 46-52, January, http://dx.doi.org/10.1016/j.tranpol.2010.05.007.

Merk, O. (2014), "Metropolitan governance of transport and land use in Chicago", *OECD Regional Development Working Papers*, No. 2014/08, OECD Publishing, Paris, http://dx.doi.org/10.1787/5jxzjs6lp65k-en.

Merk, O. et al. (2012), "Financing green urban infrastructure", *OECD Regional Development Working Papers*, No. 2012/10, OECD Publishing, Paris, http://dx.doi.org/10.1787/5k92p0c6j6r0-en.

OECD (2015), *Governing the City*, OECD Publishing, Paris, http://dx.doi.org/10.1787/9789264226500-en.

OECD (2014a), *The Cost of Air Pollution: Health Impacts of Road Transport*, OECD Publishing, Paris, http://dx.doi.org/10.1787/9789264210448-en.

OECD (2014b), *Recommendation of the Council on Effective Public Investment Across Levels of Government*, OECD, Paris, available at: www.oecd.org/gov/regional-policy/Principles-Public-Investment.pdf.

OECD (2014c), "A national strategy for cities: Taking ownership of urban policy", in: *OECD Regional Outlook 2014: Regions and Cities: Where Policies and People Meet*, OECD Publishing, Paris, http://dx.doi.org/10.1787/9789264201415-9-en.

OECD (2013a), *OECD Territorial Reviews: Puebla-Tlaxcala, Mexico 2013*, OECD Publishing, Paris, http://dx.doi.org/10.1787/9789264203464-en.

OECD (2013b), *Inventory of Estimated Budgetary and Tax Expenditures for Fossil Fuels 2013*, OECD Publishing, Paris, http://dx.doi.org/10.1787/9789264187610-en.

OECD (2013c), *Green Growth in Cities*, OECD Green Growth Studies, OECD Publishing, Paris, http://dx.doi.org/10.1787/9789264195325-en.

OECD (2013d), *Investing Together: Working Effectively across Levels of Government*, OECD Publishing, Paris, http://dx.doi.org/10.1787/9789264197022-en.

OECD (2013e), "Urbanisation and green growth in China", *OECD Regional Development Working Papers*, No. 2013/01, OECD Publishing, Paris, http://dx.doi.org/10.1787/5k49dv68n7jf-en.

OECD (2013f), *OECD Regions at a Glance 2013*, OECD Publishing, Paris, http://dx.doi.org/10.1787/reg_glance-2013-en.

OECD (2013g), *Government at a Glance 2013*, OECD Publishing, Paris, http://dx.doi.org/10.1787/gov_glance-2013-en

OECD (2012a), *Compact City Policies: A Comparative Assessment*, OECD Green Growth Studies, OECD Publishing, Paris, http://dx.doi.org/10.1787/9789264167865-en.

OECD (2012b), *Redefining Urban: A New Way to Measure Metropolitan Areas*, OECD Publishing, Paris, http://dx.doi.org/10.1787/9789264174108-en.

OECD (2012c), *OECD Environmental Outlook to 2050: The Consequences of Inaction*, OECD Publishing, Paris, http://dx.doi.org/10.1787/9789264122246-en.

OECD (2010), *Cities and Climate Change*, OECD Publishing, Paris, http://dx.doi.org/10.1787/9789264091375-en.

OECD (2009), "Investing for growth: Building innovative regions", background report for the Meeting of the Territorial Development Policy Committee (TDPC) at the Ministerial Level, 31 March 2009, www.oecd.org/regional/ministerial/42531915.pdf.

OECD/ITF (2015a), "Shifting towards low carbon mobility systems", Discussion Paper No 2015-17, May 2015 OECD/ITF, Paris.

OECD/ITF (2015b), *ITF Transport Outlook 2015*, OECD Publishing, Paris, http://dx.doi.org/10.1787/9789282107782-en.

OECD/ ITF (2015c), "Policy strategies for vehicle electrification", Discussion Paper No. 2015-16, OECD/ITF, Paris.

RMV (2014), Struktur des RMV, www.rmv.de/de/Verschiedenes/Informationen_zum_RMV/Der_RMV/Wir_ueber_uns/Struktur_des_RMV (accessed November 2014).

Smith S., and N.A. Braathen (2015), "Monetary carbon values in policy appraisal: An overview of current practice and key issues", *OECD Environment Working Papers*, OECD Publishing, Paris.

STIF (2012), "Enquête globale transport", www.iauidf.fr/fileadmin/Etudes/etude_954/Enquete_globale_transport_HD.pdf.

Transport for London (TfL) (2014), Standing Orders from 1 October 2014, www.tfl.gov.uk/cdn/static/cms/documents/tfl-standing-orders.pdf (accessed November 2014).

URBS (2014), www.urbs.curitiba.pr.gov.br (accessed November 2014).

UN (2014), *World Urbanization Prospects: The 2012 Revision*, United Nations, Department of Economic and Social Affairs, Population Division, Population Estimates and Projections Section, available at: http://esa.un.org/wpp.

Chapter 9

Strengthening incentives for sustainable land use

Sustainable land-management practices – reduced deforestation, restoring degraded land, low-carbon agricultural practices and increased carbon sequestration in soils and forests – can contribute to significant greenhouse gas emission reductions while responding to growing food demand. They could also improve the resilience of our economies to a changing climate by protecting ecosystems. Achieving this will require an integrated approach that breaks down the silos between climate change, agriculture, food security, forestry and environment policies. This chapter explores misalignments arising from the existence of environmentally harmful agricultural subsidies, the lack of valuation of ecosystem services and forest protection, and the incentives leading to food waste across the agriculture value chain.

The statistical data for Israel are supplied by and under the responsibility of the relevant Israeli authorities. The use of such data by the OECD is without prejudice to the status of the Golan Heights, East Jerusalem and Israeli settlements in the West Bank under the terms of international law.

Key messages

By 2050, land use will have to supply 60% more food than today to feed a growing population, in a way that does not harm soil, water, biodiversity, ecosystem services or the climate upon which human well-being and development depend. Yet current land use – mainly agriculture and deforestation – is responsible for around 25% of man-made global greenhouse gas (GHG) emissions.

It need not be this way. Sustainable land-management practices – reduced deforestation, restoring degraded land, better agricultural practices and increased carbon sequestration in soils and forests – could make a large contribution to the global climate change effort while delivering the productivity improvement needed to respond to growing food demands. It could also improve the resilience of our economies to a changing climate by protecting ecosystems.

Achieving this will require an integrated approach which breaks down the silos between climate change, agriculture, food security, forestry and environment policies. As a first step, this chapter explores the following specific questions:

Are agricultural support policies consistent with a low-carbon economy? Agricultural input subsidies and price support policies can reduce the environmental and climate performance of agriculture. Since 1990, concerted efforts by OECD countries have reduced the most environmentally harmful subsidies – such as for nitrogen fertilisers and fossil fuels – from over 85% of all agricultural subsidies to 49% in 2010-12. Governments should continue these efforts, while redirecting support to practices, skills and infrastructure that reduce the carbon and resource-use intensity of farming in a way that is compatible with continued productivity improvement.

Is the trade regime for agricultural products supportive of climate goals? Liberalising trade measures is important for climate change mitigation, adaptation and food security. Reducing tariffs and subsidies on agricultural products could optimise land use and reduce the overall demand for land, reducing pressure on forested areas. A well-functioning trade system would also support adaptation, compensating regional changes in productivity induced by climate change. Open trade also enhances the four pillars of food security: accessibility, availability, utilisation and stability. Yet across each dimension there is a mixture of positive and negative effects resulting from trade openness, such as concerns over import dependence for countries without a comparative advantage. These effects need to be managed.

Do policies undermine agriculture's resilience to climate change? Agriculture will likely be very hard-hit by climate change, possibly reducing yields by 25% compared to current levels in some regions, with drastic consequences for developing and emerging economies that depend on agriculture. Contradictory policy signals, as well as a general lack of capacity to access and make use of relevant climate data, can prevent farmers from integrating climate risk into their daily practices. For example, crop insurance subsidised beyond the premiums needed to see farmers take into account the climate threat could induce more risky farming practices. More support is needed to help farmers adapt to drought, flooding or other climate impacts, particularly in developing countries.

Are services provided by forests and ecosystems properly valued in economic decisions? Decisions on land-use allocation, such as between agriculture, forest and infrastructure uses, are guided by market forces, government incentives and regulations that do not always fully consider environmental costs and benefits. For instance, forests provide a wide range of climate and environmental services that are largely unpriced. Designing incentives that encourage activities with dual benefits of emissions abatement and natural asset sustainability, such as REDD+ and Payments for Ecosystems Services, has the potential to facilitate a transition to a low-carbon economy, provided that national governments help local communities address their capacity gaps.

Key messages (*cont.*)

Towards a sustainable bio-economy? An economy based on bio-energy and bio-products could drive significant low-carbon transformation. But without well-aligned policies it also risks diverting even more biological resources to purposes other than food and leading to even higher emissions. Climate-friendly bio-energy policies would imply as a start that decisions are taken on the basis of the quantification of carbon flows and other environmental impacts throughout the entire life cycle of bio-products.

Are policies joined-up enough to get to the roots of food waste? Reducing food losses and waste from field to household could help ease environmental pressures and climate impacts by improving efficiency along the agricultural supply chain. Governments need to better understand the drivers of loss and waste, and respond with co-ordinated policies that bring together the ministries of agriculture, economy, environment and health.

Land use could be key for tackling climate change

The global challenge over the coming decades will be to raise agricultural production and productivity to meet ever-growing global demand for food, feed, fibre and energy, while at the same time minimising GHG emissions and broader environmental impacts (OECD, 2014a). The relationship between land use and climate change is complex, involving important feedback loops, both positive and negative, between mitigation, adaptation and the need to increase food security and protect ecosystem services. While current land-based activities are on balance increasing emissions of GHGs, modifications could turn rural land into a net carbon sink. This chapter explores this potential in the areas of agriculture, forestry, bio-economies and food waste. Enabling these policy areas to fulfil their low-carbon potential will require an integrated approach that breaks down the silos between climate change, agriculture, food security, forestry and environment policies (OECD, 2014a).

Current land use is a major source of emissions

Land use (mainly agriculture and forestry) is responsible for around 25% of global GHG emissions. These mainly arise from deforestation and livestock, soil and nutrient management (IPCC, 2014).

Every year between 2000 and 2010, GHG emissions from agriculture represented 10-12% of all man-made emissions. The agricultural sector is the largest contributor of global anthropogenic non-CO_2 GHGs, accounting for 56% of total emissions in 2005. The main culprits are nitrous dioxide – largely derived from the application of fertilisers, manure and crop residues and the cultivation of organic soils, and methane from livestock enteric fermentation and rice cultivation. Animal agriculture is an important source of GHG emissions, and many developments in the animal sector affect potential shifts in future GHG emissions. Actual CO_2 emissions represent another 15% of total GHG emissions from agriculture, mainly from the combustion of fossil fuels in farm machinery and for powering irrigation pumps (IPCC, 2014; Figure 9.1). Between 1990 and 2011, agricultural emissions in more advanced economies decreased by 20%, while they increased by 37% in developing and emerging economies. In 2012, 76% of agriculture emissions came from developing economies (FAO, 2014a).

Figure 9.1. **The role of land use in greenhouse gas emissions and sinks, 2001-10**

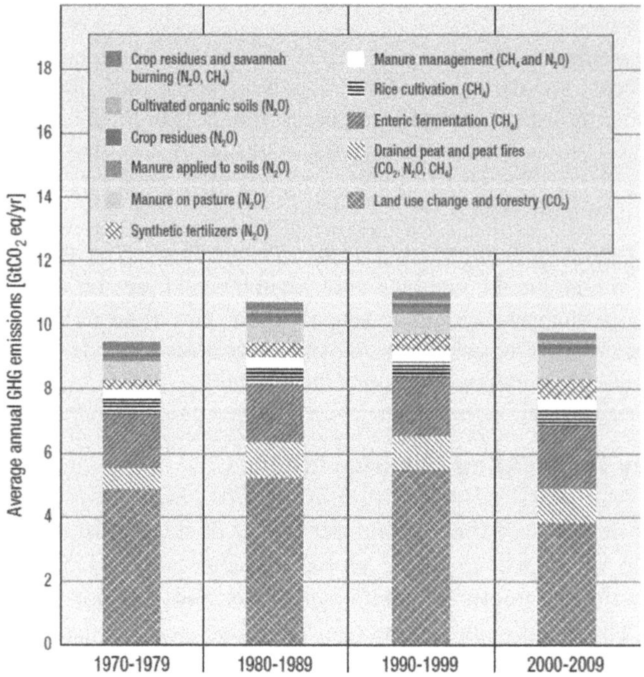

Notes: N_2O: nitrous oxide; CH_4: methane; CO_2: carbon dioxide

Source: IPCC (2014), Intergovernmental Panel on Climate Change, Working Group III, Summary for Policymakers, www.ipcc.ch.

Annual GHG flux from land use and land-use change, mainly deforestation of tropical forests and the domestic use of biomass, represents 9-11% of total man-made GHG emissions. Global emissions from deforestation declined by 3% between 2000 and 2010 (IPCC, 2014). Most net forest conversion takes place in developing countries: of total forest conversion, 54% takes place in the Americas, 6% in Africa and 15% in Asia (FAO, 2014a).

Well-managed land can make a unique contribution to the low-carbon future

Land offers unique climate change mitigation potential: it can enhance the removal of GHGs through carbon sequestration and reduce emissions through sustainable approaches to managing land and livestock. It also plays an important role in adapting to climate change. The climate mitigation and adaptation services provided by ecosystems will become increasingly important in the low-carbon transition and should be valued appropriately. Policy makers will need to adapt available instruments to their country contexts (Box 9.1).

Land offers unique climate change mitigation potential

Modelling suggests that cost-effective pathways for keeping the global average temperature increase below 2°C would require land to become a net sink of CO_2 emissions by the second half of this century (IPCC, 2014; OECD, 2012). This would compensate for hard-to-reduce emissions in other sectors. However, the role of land as a CO_2 sink relies on biological processes which are also difficult to control. Whether a particular ecosystem acts as a sink or source of GHGs may change over time. Nevertheless, the Intergovernmental

Panel on Climate Change (IPCC) estimates that land-related mitigation strategies could account for 20-60% of total cumulative GHG reductions to 2030, and 15-40% to 2100 (IPCC, 2014).

Box 9.1. Examples of policy approaches to greenhouse gas emission reductions in agriculture

There is a range of instruments governments can use to reduce greenhouse gas (GHG) emissions from agriculture. The mix of instruments will have to be adapted to specific social, institutional and economic contexts, as well as to objectives beyond climate.

- Environmental standards and regulations: Standards and rules for land management, controls on excessive use of agrochemicals and fertilisers.
- Support measures: Decouple farm support from commodity production levels and prices, remunerate provision of carbon sequestration and flood and drought control, increase investment in technologies, target environmental outcomes or production practices favourable to climate.
- Economic instruments: Impose a price on carbon and other GHG gases through taxes or trading schemes for carbon emissions.
- Trade measures: Lower tariff and non-tariff barriers on food and agriculture (bearing in mind the environmental concerns such as biodiversity and sustainable resource use), eliminate export subsidies and restrictions on agricultural products.
- Research and development: Increase public research and development on sustainable food and agriculture, promote private R&D, undertake public-private partnerships for green agricultural research.
- Development assistance: Allocate more development aid for reducing carbon emissions from agriculture.
- Information, education, training and advice: Increase public awareness for more sustainable patterns of consumption through certification and eco-labelling, particularly to reduce food waste, incorporate sustainable approaches in training, education and advice through the entire food chain.

Source: OECD (2011), *Food and Agriculture*, OECD Green Growth Studies, OECD Publishing, Paris, http://dx.doi.org/10.1787/9789264107250-en.

Land-use mitigation opportunities lie on both the supply and demand side. On the supply side, terrestrial ecosystems can help improve the global GHG balance through four main mechanisms:

1. Avoiding emissions by reducing deforestation and land degradation.
2. Reducing non-CO_2 emissions from the agricultural sector, including livestock, through improved agricultural practices (see OECD, 2015b). New technologies can reduce emissions – including improved manure storage (e.g. structures, lagoon covers), manure management (e.g. rate, timing, field incorporation), improved feed management and power generation (e.g. methane capture, manure as a biomass feedstock).

3. Enhancing CO_2 removal by increasing carbon sequestration in soils and vegetation through improved cropland and livestock management, conservation tillage practices, forestry and agroforestry, and restoration of degraded land.
4. Reducing GHG emissions in the energy, construction and transport sectors through the expanded use of low-carbon renewable and sustainable bioenergy, biofuels and bio-based materials (OECD, 2012; IEA, 2014).

On the demand side, GHG emissions could be mitigated by reducing food loss and waste, changing diets and changing behaviours in wood consumption. In the construction sector, greater use of wood from sustainably managed forests (in place of concrete, steel, etc.) could also contribute to lowering GHG emissions. However, quantitative estimates of the potential of demand-side options are few and highly uncertain (IPCC, 2014).

Beyond improving mitigation, sustainable land management practices such as conservation agriculture, intercropping and sustainable forestry can also improve the climate resilience of agriculture through the multiple benefits ecosystems provide, such as reducing erosion, building soil fertility and structure, improving water quality and buffering against drought (OECD, 2014c).

Delivering these mitigation options involves considerable challenges. A key factor will be ensuring that all costs associated with climate change are reflected in production and consumption decisions. Governments can achieve this through market-based instruments such as environmentally related taxes, charges and fees; tradable permits and subsidies for reducing pollution; or through non-market approaches such as regulations and voluntary approaches (Box 9.1). Market instruments have proven difficult to implement in agriculture due to the lack of definition of property rights, difficulties in identifying sources of pollution and agriculture's location-specific environment externalities. Regulations and payments have been easier to implement (OECD, 2013a).

There are also many potential barriers to implementation of known mitigation practices, particularly in developing countries. These include accessibility to land-use financing; poverty; institutional, ecological and technological development; diffusion and transfer barriers (IPCC, 2014). Aligning policies towards low-carbon goals would facilitate and reduce the cost of implementation of these mitigation options.

Align food production and low-carbon goals

Climate regulation through mitigation and adaptation options is just one of the many ecosystem services, goods and benefits provided by land (Figure 9.2). Land is fundamental for feeding the world's 7 billion people, supporting the livelihoods of billions of people worldwide and as a critical resource for sustainable development in many regions (OECD, 2014a). Mitigation and adaptation options in the land-use sector therefore need to be assessed, as far as possible, for their potential impact on all other services provided by land, and vice versa.

Agriculture is also vital to the livelihoods of many people, especially in developing countries, where the agricultural sector constitutes a substantial portion of gross domestic product (GDP) and of employment: 23% of GDP in Africa, 22% in Asia and 10% in Latin America, compared to around 3-4% of GDP in developed countries (World Bank, 2013). Demand for agricultural products is expected to grow by 60% by 2050 compared with 2005, in response to growing levels of per capita consumption, greater demand for protein-based diets and a world population that is expected to reach 9 billion (Alexandratos and Bruisna, 2012). The way this demand is met will have a significant impact on the climate and on the broader green growth agenda. If agriculture were to continue on a business-as-usual path to meet the rising demand for food, energy and infrastructure, natural resources would be exploited beyond their ecological carrying capacity (OECD, 2013a).

Figure 9.2. **Climate regulation is just one of many services provided by land use**

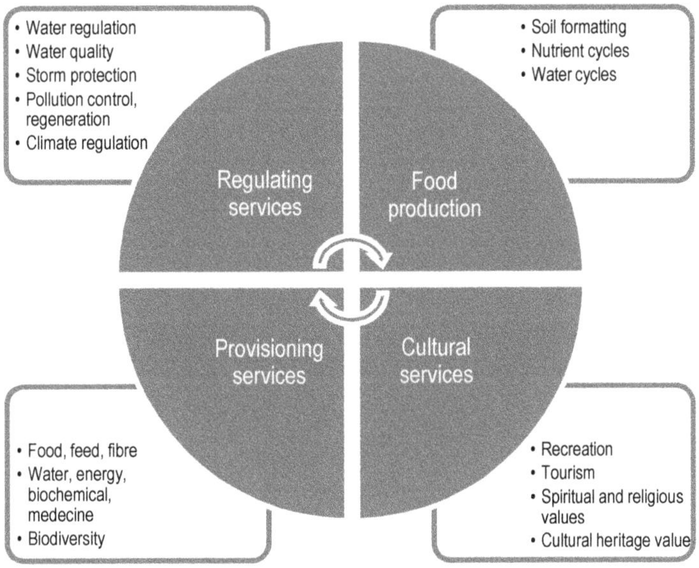

Source: Adapted from Millenium Ecosystem Assessment (2005), *Ecosystems and Human Wellbeing: Current State and Trends,* Island Press, Washington, DC.

How to reconcile food security, adaptation and mitigation agendas?

There are synergies and trade-offs between adaptation, mitigation and the broader food security agenda (OECD, 2014a). Agricultural expansion and intensification could have a negative impact on the climate agenda, which could in turn undermine efforts to increase food production.

- Changes in land use (from pasture to crop area, forest to agriculture) has an impact on GHG emissions, either directly when annual crops are grown on land that was previously forested, or indirectly when it displaces production of other commodities which are then produced elsewhere (indirect land-use change). Agricultural expansion is the primary cause of deforestation in tropical landscapes. In the absence of appropriate incentives to protect forests, the various pressures to expand agriculture – food demand, bioenergy policies – could lead to further deforestation (DeFries and Rosenweig, 2010).
- In the last decades, 70% of the increase in food production has come from agricultural intensification, using higher yielding varieties of seeds, synthetic fertilisers, irrigation and mechanisation. Projections estimate that 80% of future production increases will have to come from further intensification (FAO, 2002). While intensification on existing land can reduce deforestation and forest degradation and therefore carbon emissions, it can also involve higher emissions from chemical inputs and increased livestock densities.

- On the other hand, agriculture is itself one of the most vulnerable sectors to climate change (Box 9.2).

Box 9.2. The impact of climate change on agriculture

The agricultural sector is likely to be strongly affected by the increased variability in temperature and rainfall patterns, changes in water availability, increased frequency of extreme weather events and changes in ecosystems brought about by climate change (IPCC, 2014; OECD, 2015a).

Some regions will benefit from a changing climate, with longer growing periods in cool regions and increased carbon fertilisation. But recent analysis shows that, overall, yields will still increase globally, but less than they would have without climate change.

Climate change could cause yields of maize, wheat and rice to decline by 10%, 7% and 6% respectively on average in OECD countries, compared to a scenario without climate change. Some regions will be affected worse than others: yields of certain crops could fall by as much as 25% (for example, maize in North America and wheat in Australia).

Climate change may also have direct implications for land availability. Some areas could drop out of production altogether, while higher temperatures may see others (e.g. in Canada and in the Russian Federation) increase their arable land. The net effect of these developments is unclear, but farmers will need to continuously adapt their farming practices to changing rainfall patterns and temperatures to maintain or increase their productivity (IPCC, 2014).

Sources: IPCC (2014), Intergovernmental Panel on Climate Change, Working Group III, Summary for Policymakers, www.ipcc.ch; Ignaciuk, A. and D. Mason-D'Croz (2014), "Modelling adaptation to climate change in agriculture", *OECD Food, Agriculture and Fisheries Papers*, No. 70, OECD Publishing, Paris, http://dx.doi.org/10.1787/5jxrclljnbxq-en

Understanding the drivers of agricultural production choices, impacts on farmers and potential unintended consequences for the environment is a prerequisite to identifying misalignments between agricultural policies and climate objectives. But doing so is complex. A wide range of policies, market forces and environmental factors drive agricultural systems, practices, input use, farm outputs and the ecosystem services provided by agriculture. This in turn has implications for the state of the environment (soil, water, air, biodiversity), which then affects human activities including health, social values, agriculture, industry and urban centres (OECD, 2013b; Figure 9.3). This complexity requires a holistic and well-integrated decision-making process (Sayer et al., 2013).

In addition, these drivers vary significantly depending on the country context. Subsistence farmers in sub-Saharan Africa do not face the same constraints, nor do they have the same ability to respond to policy signals, as major global agribusinesses. Climate change impacts will also vary widely by region and will have to be addressed locally. However, there are some common areas of misalignment which should be tackled, highlighted in the sections which follow: the composition of policy support to farmers, insurance policy design, trade restriction measures for agricultural products and investment in technology and knowledge.

Move away from environmentally harmful support to agriculture

Governments support farmers and agribusinesses for several reasons: to manage the supply of agricultural commodities, to influence their cost, to supplement farmers' incomes, and to achieve other social and environmental goals such as water conservation, poverty reduction and climate change mitigation. In OECD countries, support to farmers represented USD 248 billion between 2009 and 2011 (OECD, 2013a). This involved keeping prices above market levels, making direct payments to farmers and subsidising inputs. Since the 1980s, overall support to farmers has declined from 37% to 18% in real terms and in proportion to their other receipts. While direct support to agriculture is still relatively low in emerging and developing countries, the use of subsidies to protect agricultural sectors is increasing, driven by food security and poverty agendas (Figure 9.4).

Figure 9.3. **Linkages between policies, agricultural driving forces and the state and impact of agriculture on the environment and human welfare**

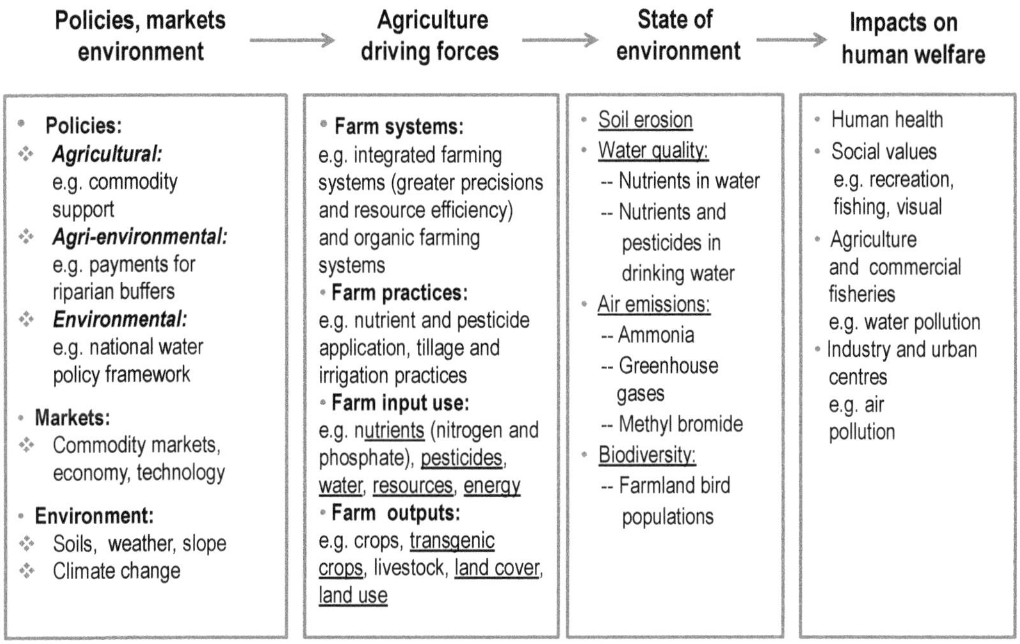

Note: The bullet points in each box are illustrative and some are interchangeable. Soil erosion, for instance, is included in the "state of environment" list, but could also be listed as a "driving force" behind soil sedimentation.

Source: OECD (2013b), *OECD Compendium of Agri-environmental Indicators*, OECD Publishing, Paris, http://dx.doi.org/10.1787/9789264186217-en.

The way governments design their support may have unintended consequences on the environment, allocation of resources and GHG intensity of agriculture (OECD, 2013a; Table 9.1).

Market price-support mechanisms and payments based on inputs create incentives for greater production. When they do not impose environmental constraints on farming practices, these are potentially the most harmful type of support for the environment. For instance, fossil fuel and fertiliser subsidies can lead to the suboptimal use of inputs and increase GHG emissions. Price support mechanisms mask market signals to farmers and can lead to inefficient use of fertilisers and pesticides, with negative climate and environmental consequences.

- Payments based on cropped land, animal numbers, historical entitlements or overall farming income are likely to be less harmful in environmental terms, as they place limits on the level of production supported (OECD, 2013a).
- Payments based on non-commodity criteria, or payments linked to constraints on resource use – such as biodiversity conservation; flood, drought and soil erosion control; and the creation of carbon sinks – are beneficial for the environment and potentially for the climate. Such payments might be made to farmers for retiring fragile

land from production, planting trees or changing tillage practices to reduce climate change impacts. These types of support payments make up a very low share of current payments across the OECD.

Figure 9.4. **Changes in producer support estimate (PSE)* in OECD and non-OECD countries, 1995-2012**

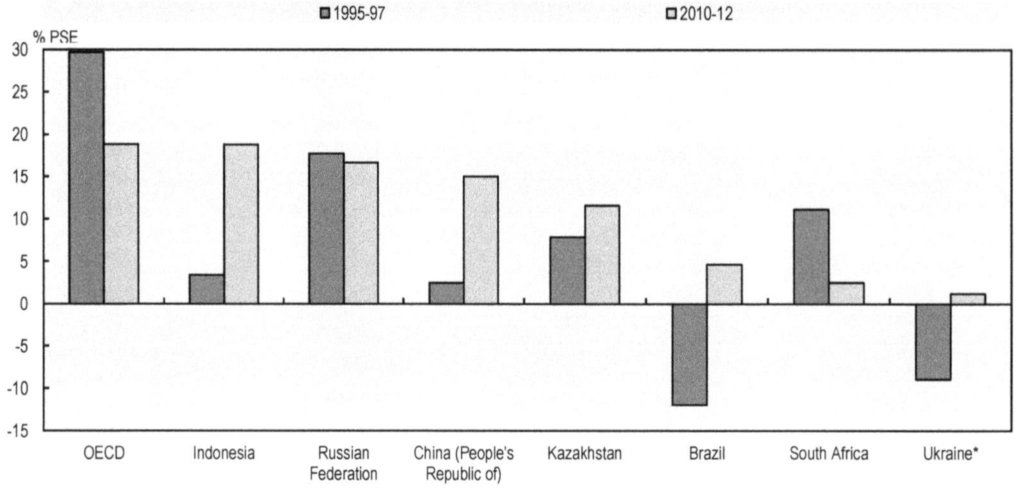

Notes: * This is based on the Producer Support Estimate (PSE), which measures support to individual farmers provided through supporting prices above market levels, by making direct budgetary payments to farmers or subsidising inputs (OECD, 2014d).

Source: OECD (2013a), *Agricultural Policy Monitoring and Evaluation 2013: OECD Countries and Emerging Economies, OECD Publishing,* Paris, http://dx.doi.org/10.1787/agr_pol-2013-en.

The actual impacts of those policies will depend on the many factors that determine farmers' overall decisions. In addition, there may be some trade-offs and conflicting objectives between the climate agenda and other environmental goals. For instance, using grazing animals to maintain a wildlife habitat can affect water quality and increase GHG emissions (OECD, 2013a).

However, low-carbon agriculture would mean shifting away from the more environmentally harmful types of support and regulations linked to climate benefits (OECD, 2013b). This would increase the effectiveness and reduce the cost of mitigation policies in agriculture, through aligning farmers' incentives and signals in the same direction (OECD, 2013a).

Since 1990, OECD countries have reduced the most environmentally harmful subsidies (those based on prices and output levels, such as for nitrogen fertilisers and fossil fuels) from over 75% of all agricultural subsidies in 1995 to 49% in 2010-12. Some domestic price support mechanisms have been replaced by direct payments based on past entitlement levels or farm income. These represent one-third of total support today, and have less production-distorting and environmental impacts. Some countries are making support increasingly conditional on environmental objectives: the EU, Switzerland and the United States link more than 50% of their support to some environmental cross-compliance.[1] In Switzerland, farms need to provide proof of ecological performance in order to be eligible for general direct payments, for example.

Table 9.1. **Potential environmental impact of Producer Support Estimate and their share in the OECD area**

% of total PSE payments

Potential environmental impact	Type of support measure	1995-97	2009-11
Potentially most harmful	Market price support	67	43
	Payments based on commodity output, without imposing environmental constraints on farming practices	3	2
	Payments based on variable input use, without imposing environmental constraints on farming practices	4	5
	Total	**74**	**50**
Potentially less harmful	Payments based on current cropped area/number of animals/receipts or income, without imposing environmental constraints on farming practices	10	5
	Payments based on historical entitlements/receipts or income, without imposing environmental constraints on farming practices	1	2
	Payments based on fixed capital formation, without imposing environmental constraints on farming practices	3	3
	Payments based on on-farm services, without imposing environmental constraints on farming practices	2	3
	Total	**16**	**14**
Potentially more beneficial	Payments subject to environmental cross-compliance[1]	5	28
Potentially most beneficial[1]	Payments based on non-commodity criteria that impose environmental constraints on farming practices	1	2
	Payments based on fixed capital formation that impose environmental constraints on farming practices	1	1
	Payments based on on-farm services that impose environmental constraints on farming practices	0	0
	Payments based on variable input use that impose environmental constraints on farming practices	0	0
	Payments based on current cropped area/number of animals/receipts or income that impose environmental constraints on farming practices	3	4
	Payments based on historical entitlements/receipts or income that impose environmental constraints on farming practices	0	1
	Payments based on commodity output that impose environmental constraints on farming practices	0	0
	Total	**5**	**8**

Note: 1. Includes payments from various PSE categories which are subject to environmental cross-compliance.

Source: OECD (2013a), *Policy Instruments to Support Green Growth in Agriculture: A Synthesis of Country Experiences*, OECD Publishing, Paris, http://dx.doi.org/10.1787/9789264203525-en.

The emergence of agri-environmental and environmental policies since the early 1990s – e.g. payments to remove land from production or to provide financial incentives to encourage certain types of production practices considered beneficial for the environment – has encouraged farmers in many OECD countries to use fewer chemicals on their land (OECD, 2013b). These policies have been accompanied by an overall decrease of GHG emissions from agriculture and an improvement in the environmental efficiency of agricultural GHG emissions (i.e. GHG emissions have risen more slowly than increases in agricultural production) (Figure 9.5).

However, reform has been uneven across OECD countries. While some countries have taken clear steps to decouple support from output and price levels, others have not yet begun to address the problem. Further, the most environmentally beneficial type of support, agri-environmental schemes, only represent 2% of total agricultural support in

OECD countries (Table 9.1 and Figure 9.6). The use of cross-compliance mechanisms is better than providing direct income support without environmental conditionality attached, but is not as cost-efficient as targeted measures.

Figure 9.5. **Greenhouse gas emissions, GDP and productivity for agriculture in the OECD area, 1990-2010**

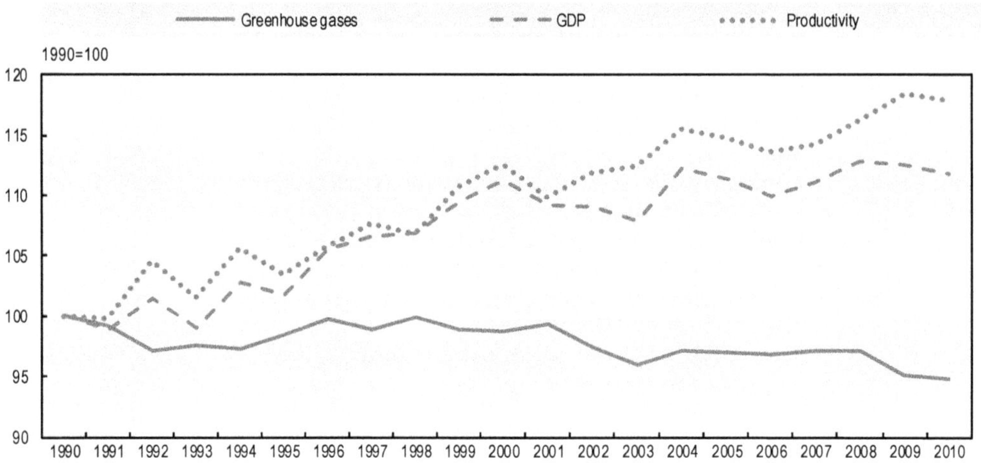

Note: Excluding LULUCF (land use, land-use change and forestry).

Source: OECD (2014d), *Agricultural Policy Monitoring and Evaluation 2014: OECD Countries*, OECD Publishing, Paris, http://dx.doi.org/10.1787/agr_pol-2014-en.

Figure 9.6. **Evolution of producer support by potential environmental impact in the OECD area, 1990-2012, in USD million**

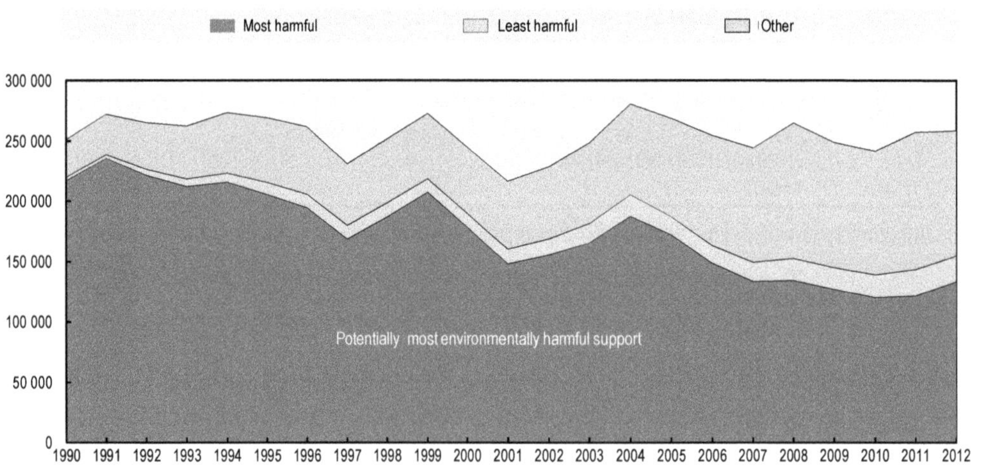

Note: The source of the classification is OECD (2013a), *Policy Instruments to Support Green Growth in Agriculture: A Synthesis of Country Experiences*, OECD Publishing, Paris, http://dx.doi.org/10.1787/9789264203525-en.

Source: OECD (2014h), *Green Growth Indicators for Agriculture: A Preliminary Assessment*, OECD Green Growth Studies, OECD Publishing, Paris, http://dx.doi.org/10.1787/9789264223202-en.

In emerging economies, agricultural support generally relies heavily on production-distorting policies such as market price support and input subsidies. Net exporters such as Brazil and South Africa provide limited levels of subsidies, but of the more distortive types (i.e. price support and input subsidies). In the People's Republic of China (hereafter "China") and Indonesia, grain prices (wheat and rice, respectively) are regulated and local producers are either supported or taxed depending on the levels of international prices. Both countries provide input subsidy support, and China provides direct payments by unit of land (OECD, 2014d). India provided roughly USD 28 billion in input subsidies to nitrogenous fertilisers and electricity for pumping agricultural water in 2010 (GCEC, 2014). Fertiliser subsidies in India alone represented USD 8 billion in 2007, or 0.8% of GDP (Von Lampe et al., 2014). Removing such subsidies would allow more efficient input use and reduce associated GHG emissions (GCEC, 2014).

The use of input subsidies has seen a resurgence in least-developed countries, in Africa in particular, driven by the belief that subsidising the cost of inputs would create a virtuous circle of higher yields and less hunger and poverty. However, as in any other country, the costs and benefits of input subsidies must be compared with conventional best practices of addressing market failures directly, such as using social policies (e.g. cash transfers, support to investment and risk-mitigation tools) to tackle poverty and food insecurity (Brooks, 2014).

Avoid trade restrictions for more efficient and resilient global food markets

Liberalising trade measures is important for climate change mitigation, adaptation and food security. However, in the absence of adequately priced externalities, the potentially negative side-effects of open trade (see Chapter 5) need to be managed.

Trade and mitigation

Reducing tariffs and subsidies on agricultural products could optimise land use and reduce the overall demand for land, reducing pressure on forested areas. As these subsidies distort price signals, they could create misallocation of land uses: e.g. land currently used for wheat might be more suitable for livestock (OECD, 2013a).

However, as GHG externalities are not properly priced in the market, opening trade barriers could, in some cases, mean a geographical shift of agricultural production. Without appropriate measures, the result could be additional GHG emissions (if the shift implies additional deforestation in tropical zones, for instance).

In a globalised world, national strategies for forest protection may also have unintended consequences. In particular, national conservation measures to protect forests and promote reforestation could lead to deforestation in other areas via the trade in agricultural and wood products, undermining the effectiveness of those measures (Meyfroidt and Lambin, 2011). Within a given country, this could be ameliorated by a national land-use plan which includes restoration of degraded land and increased productivity of existing land under cultivation. Without harmonised sustainability standards, the availability of cheap and sometimes unsustainable timber from abroad can also lower domestic prices, providing a barrier to investment in reforestation and sustainable forest management at home. In Viet Nam, policies restricting forest exploitation combined with the rapid development of the wood-processing industry have shifted forest extraction to the neighbouring countries of Cambodia and Laos. Evidence suggests that in the seven developing countries whose forests are increasing overall, additional emissions from land-use change embodied in their net wood trade offset 74% of the emission captured in reforested areas, 52% when both agriculture and forestry sectors are included (Meyfroidt and Lambin, 2011).

As local land-use decisions are increasingly driven by trends in distant markets, governments should develop appropriate regulations and economic incentives to ensure that trade in wood and agricultural products yields benefits for natural ecosystems, the climate and local communities (Lambin et al., 2014). The EU has sought to restrict the flow of illegally harvested timber into the EU through the EU Timber Regulation and its bilateral agreement programme with developing countries (EC, 2013). This legislation requires importers of wood products to demonstrate that the timber was legally harvested in its country of origin (EU, 2013).

Voluntary and incentive-based instruments, often designed by private actors, could also have an important role to play, provided that they are well-designed, easy to implement and come with monitoring and verification mechanisms (Lambin et al., 2014). They include eco-certification schemes, geographical indications, commodity roundtables, and moratoria on certain product types.

Trade and adaptation

A well-functioning trade system would also support adaptation, compensating regional changes in productivity. Open trade is an important vehicle to fully reflect shifting comparative advantages due to climate change, while also pooling the risk so that yield losses in a given region can be offset through imports (OECD, 2014c). However, in an attempt to improve food security and adjust to price volatility and food shortages linked to weather-related events, some developing and emerging economies have recently shown a renewed willingness to implement trade-distorting policies such as export restrictions and export taxes (Figure 9.7).

Figure 9.7. **Number of countries with export restrictions on food staples, 2007-13**

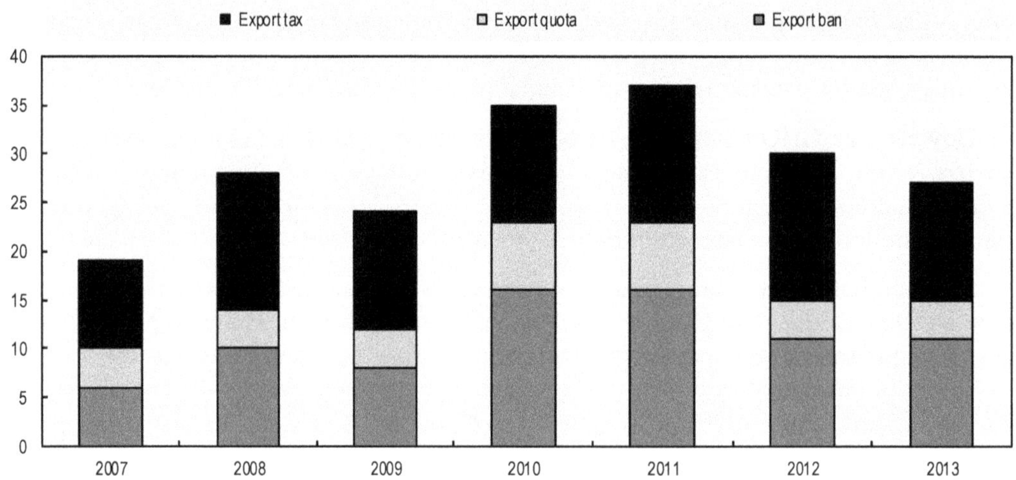

Notes: Established at the request of the agriculture ministers of the G20 in 2011, the Agricultural Market Information System (AMIS) is an interagency platform to enhance food market transparency and encourage the co-ordination of policy action in response to market uncertainty. Overall, AMIS countries represent 80-90% of global production, consumption and trade volumes.

Source: AMIS policy database.

This approach runs counter to the need to adapt to climate change, which will actually require greater land-use flexibility. The potential for "transformational adaptation" (shifting to land use more suitable to the changing climate) will be limited by these bans on exports or imports of products. With a projected increase in the incidence and magnitude of severe weather events, the implementation of these trade-restrictive policies raises future risks (Clark et al., 2012).

In addition, restrictive trade policies are not the most efficient way to address the needs of consumers and producers. Targeted social programmes, including cash transfers, favouring investment and developing risk-mitigation tools, can improve farmers' resilience to international price shocks. Safety nets (food vouchers or cash transfers) targeted at poor households are more efficient and effective than price support measures, which mainly benefit large farmers and landowners (OECD, 2013d).

Trade and food security

Open markets are also instrumental for raising output and income, enabling production to be located in areas where resources can be used most efficiently (OECD, 2013d) and improving food security. On balance, open trade enhances the four pillars of food security (Box 9.3).

Box 9.3. **The dimensions of food security**

The 1996 Food and Agriculture Organization (FAO) World Food Summit stated that: "food security exists when all people at all times have physical and economic access to sufficient, safe and nutritious food that meets their dietary needs and food preferences for an active and healthy life". People are food secure when sufficient food is **available**, when they have **access** to it, and when it is well **utilised**. The fourth element of food security is **stability** of these three dimensions over time.

Ensuring food security is not simply a question of producing more food. It requires tackling availability, accessibility, utilisation and stability at the same time. The main obstacle to food security is poverty, as low incomes constrain people's access to food. As half of the world's poor today depend directly or indirectly on agriculture, agricultural reforms need to consider their impact on farmers' incomes. Increasing availability requires increasing productivity, improving the management of supply chains and reorienting demand. Stability requires less volatile markets (OECD, 2013d).

Source: World Food Summit (1996), *Rome Declaration on World Food Security*, Food and Agriculture Organization, Rome, available at: www.fao.org/docrep/003/w3613e/w3613e00.HTM.

Yet across each dimension there is a mix of positive and negative effects resulting from trade openness, such as concerns over import independence for countries without a comparative advantage. These effects need to be managed (Brooks and Mathews, 2015):

- In terms of **availability**, open trade would allow products to flow from surplus to deficit areas; however, there are concerns about import dependency for countries without a comparative advantage in food production. There is also a risk that food supplies could be interrupted.
- With regard to **access**, open trade can reduce overall commodity prices. However, there are concerns about the impact of greater openness on the incomes of those who were formerly protected.

- Rising incomes may increase the **utilisation** of food, but also contribute to a "nutrition transition", meaning a shift to diets with more animal products, under which not all effects are positive.
- Regarding **stability of supply**, open markets reduce the risks associated with poor domestic harvests, particularly in a context of climate change, as risks are pooled in the international food market, but they make international instability a more relevant issue.

While governments should encourage more open markets and the removal of trade-restrictive measures, they should also properly assess the positive and negative of trade openness, and specifically address any negative trade-offs (Brooks and Mathews, 2015).

Remove insurance subsidies that discourage climate-resilient behaviours

With increased exposure to market forces and fluctuations, risk management has become an essential tool to help farmers better adapt and be more resilient to shocks. Risk management practices influence the ability of the agriculture sector to both mitigate GHG emissions and adapt to a changing climate. Well-designed risk management practices should lead to more efficient use of resources, higher productivity and potentially lower GHG emissions.

One risk management practice is to insure crops against catastrophic events, especially drought. However, markets for this type of insurance rarely develop without significant government support in OECD countries (notably through crop insurance subsidies and public reinsurance of last resort). Subsidised insurance is one way of providing disaster assistance to farmers. However, insurance premiums that do not adequately reflect the underlying risks can impede climate change adaptation or even promote maladaptation, for instance by reducing incentives to change to more resilient crops or by inducing farming in risky locations or with risky practices (OECD, 2015a).

Instead, governments should create an enabling environment that allows farmers to make timely, well-informed and efficient responses to the risk they are facing (OECD, 2011; 2015a). This means:

- Taking a holistic approach to risk management rather than focusing on a single source of risk, such as prices: Risks in agriculture are interconnected and compound or offset each other. Prices of inputs and outputs can move in the same direction, and lower yields often mean higher prices in a market-driven economy.
- Defining boundaries between layers of risks: While government intervention can be justified for catastrophic risk, coping with normal business risks should be left to the farmer. Government interventions should not crowd out private insurance schemes for risks such as those related to weather fluctuations, price movements or yield variations. Efficient insurance systems should be based as far as possible on risk-based premiums to provide incentives for better adapted production choices that limit the exposure and vulnerability of farm systems to climate change impacts, such as droughts and floods (OECD, 2015d). Hence there is a need to define boundaries between layers of risks as a prerequisite to government intervention. Innovative tools such as catastrophe bonds and weather index-based insurance could also be developed as a complement to market and policy responses.
- Facilitating information, regulation and training on climate change risks: Insurance markets are often under-developed due to information asymmetries between insurance

companies and farmers. Developing databases and information-sharing mechanisms will increase the options for private sector intervention. For example, governments could invest in weather forecast technologies and early-warning systems to support the emergence of index insurance covering risky events. The World Bank and the International Finance Corporation have contributed to weather risk insurance schemes in India and Africa (OECD, 2014c).

Increase investment in technological innovation and knowledge

The level of technological development and innovation in agriculture has a direct impact on its capacity to produce adequate supplies of food and feed in an environmentally sound manner (OECD, 2014b). While innovation played a significant role in the increased gains of productivity of the second half of the 20th century, continuing to focus on productivity alone may lead to natural resource depletion. A paradigm shift is needed from increased productivity to sustainably increased productivity. Sometimes called the "double Green Revolution" (OECD, 2013b), this will rely on the emergence of new technologies and the adoption of innovative farming practices that will encourage economic efficiency and environmental performance. There is a broad consensus that this transformation will not be possible without considerable investment in agricultural innovation systems. Agricultural innovation is important in two respects: disseminating improved agricultural inputs and developing new techniques, practices and technologies that address new challenges being faced by agriculture, such as climate change (OECD, 2014c; 2015a). Hence, innovation is not only about technological improvements but also about education, training and organisational improvements. Farmers will only adopt new technologies if the investment is expected to be profitable, and if they have adequate information, skills and training (OECD, 2013b).

Most OECD governments need to scale up their public budgets for agriculture-related research and development (R&D). Only 4% of public and private R&D spending in OECD countries is oriented towards agriculture, despite the importance of R&D in food security agendas. R&D support represents a very small share of total support for agriculture – around 2% in the OECD area in 2009-11. Even though government funding for R&D is permitted under international trade agreements, it accounts for just a small share of total support to agriculture (Figure 9.8) (OECD, 2013b).

In a context of limited government financial resources, encouraging the participation of the private sector in green agricultural R&D, and in agriculture more generally (Box 9.4), is essential (OECD, 2013a). This will not only increase funding opportunities, but will also ensure that research programmes meet private sector demand (OECD, 2014f). Governments can promote private R&D investments through targeted support such as tax credits and public-private partnerships (OECD, 2013a). Successful examples of private sector involvement in R&D include the Australian Rural Research and Development Corporation (funded through compulsory levies on industries and matching public contributions). Similar co-operative research programmes are in place in Canada, Denmark, the Netherlands, the United Kingdom and the United States. Tanzania has managed to heavily involve the private sector in R&D for export crops, through the Tea Research Institute of Tanzania, and institutes dedicated to tobacco and coffee (OECD, 2014f). Research costs can be reduced through greater R&D regional co-operation and with increased collaboration on training, technology transfer and knowledge-sharing across borders. This is the objective of the First Agricultural Productivity Programme for Eastern Africa, 2009-15 (ibid.).

Figure 9.8. **Government expenditures on R&D as a share of total support to agriculture**

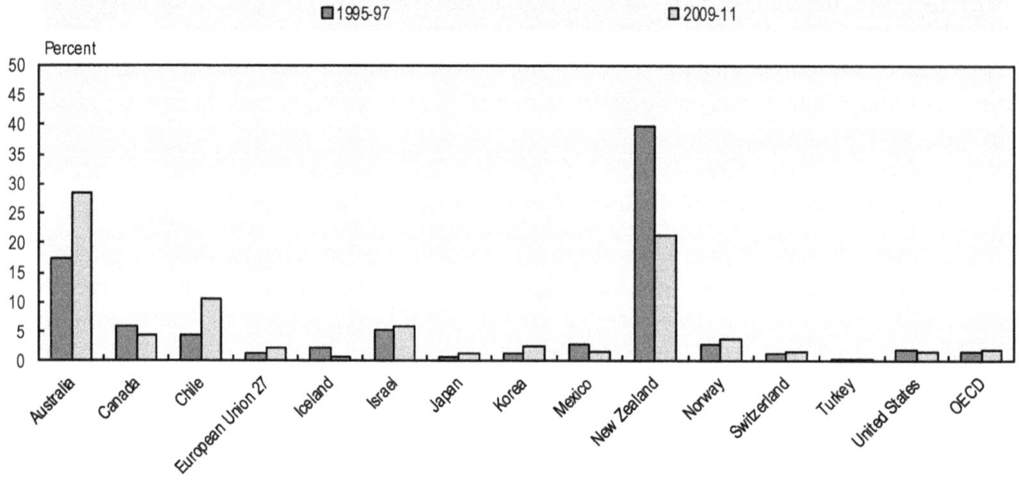

Source: OECD Producer and Consumer Support Estimates Database (2012), www.oecd.org/tad/agricultural-policies/producerandconsumersupportestimatesdatabase.htm#browsers.

Investing in knowledge is also commonly perceived to be a key driver of innovation processes in agriculture. Agricultural advisory services, training and extension initiatives play an important role in supporting green growth in agriculture and enabling farmers to meet new challenges, such as adopting environmentally sustainable farming practices and improving competitiveness (OECD, 2015c). Better trained and educated farmers are more likely to use resource-efficient practices and reduce the GHG emissions resulting from their farming practices. Learning by doing is the main form of training for the majority of farmers across OECD countries (OECD, 2015c). In the European Union, less than 10% of farm managers have completed full agricultural training. An important structural change is also the general ageing of farming population due to a low exit rate from farming of elderly farmers and a low entrance rate of young farmers. Attracting new entrants is a challenge for governments (OECD, 2014b). In developing countries, where poor and smallholder farmers are more numerous, efforts are needed to bridge the gap between the creation of innovation and its adoption.

OECD work indicates that predictable, flexible and stringent environmental policies often lead to higher investment in innovation (OECD, 2013b; see Chapter 3). On the other hand, poorly designed regulations can limit the effectiveness of investment in agricultural R&D. In the European Union, impact assessments of new regulatory proposals are now mandatory, and take into consideration the promotion of greater productivity and resource efficiency and their impacts on R&D and on intellectual property rights (OECD, 2014b). Defining an intellectual property rights regime conducive to green innovation is particularly challenging: on the one hand, governments have to provide strong protection to enable a secure environment for foreign and local firms to invest in innovation. On the other hand, they should ensure that small investors can afford valuable technologies (OECD, 2014c; 2014b).

Box 9.4. OECD Policy Framework for Investment in Agriculture

The FAO estimates that an average net investment of USD 83 billion a year will be necessary to raise agricultural production by 60% and feed the global population of more than 9 billion expected by 2050. This represents a 50% increase on current investments, and does not include the investment needed in infrastructure, storage facilities, market development or R&D.

The OECD has developed a Policy Framework for Investment in Agriculture (PFIA) to support countries in evaluating and designing policies to mobilise private investment in agriculture for steady economic growth and sustainable development. Attracting private investment in agriculture relies on a wide set of supply-side policies and on sector-specific public goods. A coherent policy framework is an essential component of an attractive investment environment for all investors, be they domestic or foreign, small or large. The PFIA is a flexible tool proposing ten policy areas to be considered by any government interested in creating an attractive environment for investors and in enhancing the development benefits of agricultural investment.

Figure 9.9. **Policy Framework for Investment in Agriculture**

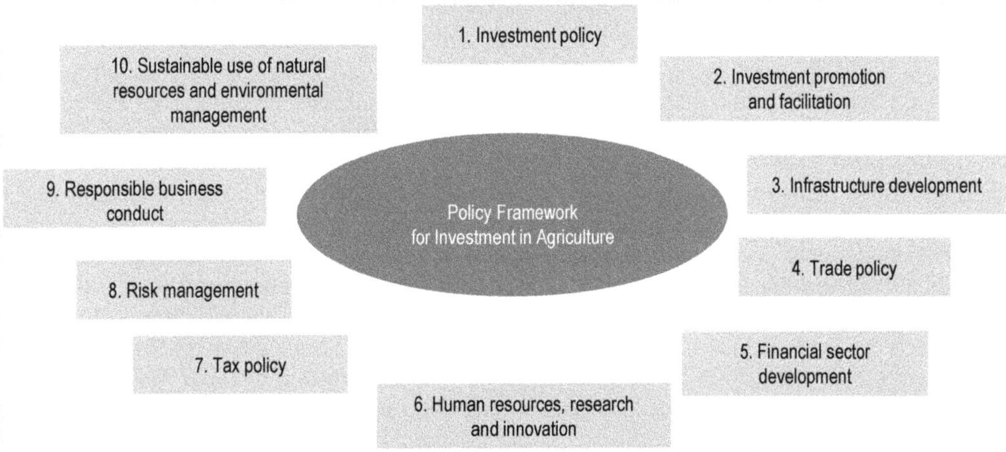

In 2014, governments from around the world approved the *Principles for Responsible Investment in Agriculture and Food Systems*, aimed at assuring that cross-border and corporate investment flows lead to improved food security and sustainability and respect the rights of farm and food workers. The principles are voluntary and non-binding, but represent the first time that governments, the private sector, civil society organisations, UN agencies and development banks, foundations, research institutions and academia have been able to come together and agree on what constitutes responsible investment in agriculture and food systems.

Sources: OECD (2014f), *Policy Framework for Investment in Agriculture*, OECD Publishing, Paris, http://dx.doi.org/10.1787/9789264212725-en; FAO (2014c), *Principles for Responsible Investment in Agriculture and Food Systems*, Food and Agriculture Organization, Rome, available at: www.fao.org/3/a-ml291e.pdf.

While technology and innovation offer huge potential, new technology can also generate additional environmental pressures or put a strain on material availability. It often involves new or substitute materials whose consequences may not yet be known (OECD, 2013b).

Break down the silos between adaptation, mitigation and food security policies

There are important synergies and trade-offs among mitigation, adaptation and food security, as well as the conservation of natural resources such as water and terrestrial and aquatic biodiversity (OECD, 2014a). Realising them requires co-ordination. Unfortunately, there is currently a lack of co-ordination between sectoral agricultural development plans and climate change policies. This can lead to inefficient use of funding, and could block the integrated management required to address climate issues while also ensuring productivity improvement and provision of food (Sayer et al., 2013).

Climate-smart agriculture, through the promotion of production systems that sustainably increase agricultural productivity and incomes while adapting and building resilience to a changing climate and reducing or removing GHG emissions, could be part of the solution (FAO, 2013; OECD, 2014a). Several global initiatives have emerged around this concept, including the Global Alliance on Climate-Smart Agriculture (Box 9.5) and the Global Research Alliance on Agricultural Greenhouse Gases.[2]

Box 9.5. The Global Alliance on Climate-Smart Agriculture

The Global Alliance on Climate-Smart Agriculture was launched in September 2014 by the United Nations Secretary General, the World Bank and the United Nations Food and Agriculture Organization (FAO), together with 46 countries including 9 OECD countries and multiple other institutions (Climate Summit, 2014). It aims to pursue the "triple win" of sustainable and equitable increases in agricultural productivity and incomes; greater resilience of food systems and farming livelihoods; and reduction and/or removal of greenhouse gas emissions from agriculture (including the relationship between agriculture and ecosystems), where possible. The Alliance seeks to improve people's food security and nutrition in the face of climate change. It will help governments, farmers, scientists, businesses and civil society, as well as regional unions and international organisations, to adjust agricultural, forestry and fisheries practices, food systems and social policies so that they take better account of climate change and the efficient use of natural resources. Members will work toward sustainable increases in the productivity of food systems, by a sustainable management of natural resources – including soil, water and biodiversity, the adaptation of livelihoods threatened by climate change, and agricultural practices that contribute to reduced emissions and less deforestation/land degradation. The Alliance will enable governments and other stakeholders to make these transformations in ways that bridge traditional sectoral, organisational and public/private boundaries. Other initiatives contribute to the same agenda, such as the Global Research Alliance on Agricultural Greenhouse Gases.

Source: Climate Summit (2014), "Global Alliance for Climate-Smart Agriculture: Action plan", available at: www.un.org/climatechange/summit/wp-content/uploads/sites/2/2014/09/AGRICULTURE-Action-Plan.pdf.

The concept of "landscape approaches" is also increasingly gaining traction. This "seeks to provide tools and concepts for allocating and managing land to achieve social, economic and environmental objectives in areas where agriculture, mining and other productive land uses compete with environmental and biodiversity goals" (Sayer et al., 2013; Box 9.6). Landscape approaches are location-specific strategies for sustainable land use which take a whole-landscape, multi-sector perspective and are based on principles of adaptive management and stakeholder engagement (DeFries and Rosenweig, 2010). They could help break down the silos between different stakeholders and objectives, and help solve the trade-offs.

Box 9.6. Ten principles of a landscape approach

Principle 1: Continual learning and adaptive management. Landscape processes are dynamic. Despite the underlying uncertainties in causes and effects, changes in landscape attributes must inform decision making. Learning from outcomes can improve management.

Principle 2: Common concern entry point. Solutions to problems need to be built on shared negotiation processes based on trust. Trust emerges when objectives and values are shared. However, stakeholders have different values, beliefs and objectives. Totally aligned objectives are unlikely, costly to establish or devoid of immediate significance. Identifying immediate ways forward through addressing simpler short-term objectives can begin to build trust.

Principle 3: Multiple scales. Numerous system influences and feedbacks affect management outcomes, but these impacts unfold under the influence of a diverse range of external influences and constraints.

Principle 4: Multifunctionality. Landscapes and their components have multiple uses and purposes, each of which is valued in different ways by different stakeholders. Tradeoffs exist among the differing landscape uses and need to be reconciled.

Principle 5: Multiple stakeholders. Multiple stakeholders frame and express objectives in different ways (Principle 2). Failure to engage stakeholders in an equitable manner in decision-making processes will lead to suboptimal, and sometimes unethical, outcomes. All stakeholders should be recognised, even though efficient pursuit of negotiated solutions may involve only a subset of stakeholders. Solutions should encompass a fair distribution of benefits and incentives.

Principle 6: Negotiated and transparent change logic. Trust among stakeholders is a basis for good management and is needed to avoid or resolve conflicts. Transparency is the basis of trust (Principle 2). Transparency is achieved through a mutually understood and negotiated process of change and is helped by good governance.

Principle 7: Clarification of rights and responsibilities. Rules on resource access and land use shape social and conservation outcomes and need to be clear as a basis for good management. Access to a fair justice system allows for conflict resolution and recourse.

Principle 8: Participatory and user-friendly monitoring. Information can be derived from multiple sources. To facilitate shared learning, information needs to be widely accessible. Systems that integrate different kinds of information need to be developed.

Principle 9: Resilience. Wholesale unplanned system changes are usually detrimental and undesirable. System-level resilience can be increased through an active recognition of threats and vulnerabilities. Actions need to be promoted that address threats and that allow recovery after perturbation through improving capacity to resist and respond.

Principle 10: Strengthened stakeholder capacity. People require the ability to participate effectively and to accept various roles and responsibilities. Such participation presupposes certain skills and abilities (social, cultural, financial).

Source: Sayer, J. et al. (2013), "Ten principles for a landscape approach to reconciling agriculture, conservation, and other competing land uses", *Proceedings of the National Academy of Sciences*, Vol. 110, No. 21, pp. 8 349-8 356, http://dx.doi.org/10.1073/pnas.1210595110

Sustainable forest management should be at the core of a low-carbon, resilient society

Degradation and loss of the world's tropical forests are together responsible for about 10% of net global GHG emissions (GCEC, 2014; IPCC, 2014). Tackling the destruction of tropical forests is therefore fundamental to any concerted effort to combat climate change. Forests contain vast carbon stocks – destroying them emits CO_2 into the atmosphere and removes an important carbon sink. Forests are also important from an adaptation perspective, as they prevent soil erosion, protect watersheds, regulate water flow and generate rainfall. Forests provide other ecosystem services such as the regulation of diseases, livelihoods (providing jobs and local employment), food and nutrient cycling.

Protecting forests is therefore vital for the continued provision of essential life-sustaining services and for the transition to a low-carbon economy. Over the past 50 years in OECD countries, the area of forests and wooded land has remained stable or has slightly increased. However, on a global scale, forest cover is decreasing due to continued deforestation in tropical countries (IPCC, 2014).

Globally, agriculture and poorly managed timber extraction are still the most important driving forces behind deforestation, with commercial and subsistence activities accounting for 40% and 33% respectively. Mining, infrastructure and urban expansion are responsible for the remainder (IPCC, 2014). During the 1980s and 1990s, rainforests were the primary sources of new agricultural land throughout the tropics. Lower production costs and limited environmental regulations allowed forest-rich tropical countries such as Brazil, Indonesia and Malaysia to respond quickly to increased demand for crops, particularly sugar cane, soybeans and palm oil. More than 80% of new agricultural land came from intact forests (Gibbs et al., 2009). Though the rate of deforestation is decreasing, it is still a major source of GHG emissions globally.

The global restoration and protection of forests is required for the low-carbon transition. This requires simultaneous alignment among several policy areas:

- domestic regulatory instruments that protect forested areas (protected areas, zoning and land-use restrictions)
- domestic agricultural and forestry policies that promote technological innovation and the sustainable use of land
- macroeconomic, trade and fiscal policies, and property law that indirectly affect land use
- policies that enable consumers to have an increasing influence on the future of the forest.

The following sections highlight misalignments in policies that indirectly affect forest land use. Other obstacles stand in the way of limiting forest degradation and deforestation, such as corruption and vested interests around forest exploitation. These governance failures call for increased policy intervention but are not the subject of this chapter (GCEC, 2014).

Financial incentives for forest countries are not yet compelling enough to drive forest-friendly development

As with agriculture, preserving forest areas requires first and foremost that land is priced in a way that reflects its true social and environmental costs and benefits. Failing to adequately value these services can lead to rapid and large misallocations of land use.

Ecosystems accounting and the concept of valuing natural capital are useful approaches for pricing forest and land-use externalities. They can help countries to assess the value of competing land uses, providing a basis for cost-benefit analysis and help governments take into account distributional impacts when designing green growth policies (OECD, 2010). Several initiatives exist at the global level: the UN Statistical Commission has developed an internationally agreed method, the System for Environmental Accounts (SEEA), and the World Bank has launched a global partnership to help countries implement natural capital accounting. In practice, however, these methods are complex to implement and often require a tailor-made approach.

REDD+ (Reduction in Emissions from Deforestation and Forest Degradation) has the potential to facilitate a transition to a low-carbon economy.[3] It is a framework through which developing countries are rewarded financially for forest protection actions which lead to a measured decrease in the conversion of forests to other land uses. The financial rewards they receive depends on the extent of the emissions reductions they achieve. Adjustments to the framework in 2007 mean that REDD+ now takes a more comprehensive approach to include the role of conservation, sustainable management of forests and the enhancement of forest carbon stocks in developing countries.

REDD+ has driven significant progress in recent years, including developing an appropriate international policy framework,[4] building the necessary capacity for implementing REDD+ programmes and piloting performance-based REDD+ (e.g. the World Bank's Forest Carbon Partnership Facility Readiness Fund and Carbon Fund). As a result, payments for verified emissions reductions are increasing, and REDD+ offsets now account for 80% of all transactions of forest carbon offsets (Ecosystem Marketplace, 2014).

However, experience to date suggests the need for a better alignment of international and national support to the scheme. Human and technical capacity at the local level is often still lacking, despite significant investment. While some forest nations may be self-motivated to take action to protect forests, for a range of economic, environmental and social reasons, realising the full mitigation potential of REDD+ will require a significant increase in financial incentives, and for interventions to be designed in a way that reinforces and supports sustainable approaches in those countries, including agricultural subsidies, rural development programmes and tax policies (Fishbein and Lee, 2015).

Another approach – payments for ecosystem services (PES) – could be instrumental in reflecting the true value of ecosystem services. PES often provide uniform per-hectare payments for protecting an important natural habitat. Reverse-auction mechanisms, as used in Australia for old-growth forests, in Indonesia to reduce soil erosion and in the United States to improve agri-environment practices, can help to improve the cost-effectiveness of PES. For example, the use of a reverse auction by the Tasmanian Forest Conservation Fund resulted in a 52% gain in cost-effectiveness compared with traditional approaches on a first-come, first-served approach (OECD, 2010). The OECD has developed a list of 12 key criteria essential for increasing the cost effectiveness of PES (Box 9.7).

Ensure clear land rights

Appropriate regulations, such as land-use zoning policies, are important for putting countries on the path to net reforestation. But they have limits, including the need for strong enforcement capacity. This is often lacking in least-developed economies where land-use zoning policies also often conflict with informal customary land tenure systems.

Box 9.7. Towards cost-effective payments for ecosystem services (PES)

1. Remove perverse incentives such as environmentally harmful subsidies.
2. Clearly define property rights to empower communities.
3. Clearly define PES goals and objectives to enhance transparency and avoid *ad hoc* political influence.
4. Develop a robust monitoring and reporting framework to assess the performance of PES.
5. Identify buyers and ensure sufficient and long-term sources of financing.
6. Identify sellers and target ecosystem service benefits.
7. Establish baselines and target payments at ecosystems that are at risk of loss or to enhance their provision.
8. Differentiate payments based on the opportunity costs of service provision.
9. Consider bundling multiple ecosystem services.
10. Address leakages.
11. Ensure permanence.
12. Deliver performance-based payments and ensure adequate enforcement.

Source: OECD (2010), *Paying for Biodiversity: Enhancing the Cost-Effectiveness of Payments for Ecosystem Services*, OECD Publishing, Paris, http://dx.doi.org/10.1787/9789264090279-en.

Communities have more incentive to manage natural capital sustainably, in particular forests and fisheries, if their ownership rights are properly defined, monitored and enforced (Ostrom, 1990). Land rights, including collective land rights, tend to encourage better land management even if there is no profit. Poorly defined property rights and weak enforcement in such cases leads to open-access exploitation (OECD, 2014e). Community resource management requires support mechanisms at the national level, such as education and training.

Establishing well-functioning land registration systems can reconcile the interests of large investors and smallholders provided that they are developed on the basis of inclusive consultation processes and consider the social, cultural and environmental values of land as well as its economic value (OECD, 2014g).

The observance and respect of land policy in each country is important for avoiding conflict over land-based investment, especially in less developed countries. The *Voluntary Guidelines on Responsible Governance of Tenure of Land, Fisheries and Forests in the Context of National Food Security* (FAO, 2012), endorsed by the Committee on World Food Security in 2012, guide governments in establishing a well-defined land-rights system.

Moving towards a bio-economy consistent with climate and development objectives

Several countries have published their plans for the development of a future bio-economy, an economy in which bio-based materials and production techniques will contribute significantly to economic and environmental sustainability. Such plans typically involve building a bio-based production industry in which fuels, energy and materials such as chemicals and plastics, usually generated from fossil resources such as oil and natural gas, are incrementally replaced by equivalent or novel products generated from renewable resources (OECD, 2014g). The bio-economy is often associated with the concept of a recycling economy that aims at reusing raw materials. Bioenergy and biofuels policies are part of a bio-economy.

A shift to a bio-economy could have significant transformative potential for the low-carbon agenda and deliver the "negative" emissions required to meet the 2°C target (IPCC, 2014). For example, sound bio-energy choices, sourced from afforestation on marginal agricultural land or sustainably managed forest and combined with carbon capture and storage, can contribute to a net sequestration of carbon. But there may be some potential unintended policy conflicts between this and the need for food security and environmental sustainability. Without well-aligned policies, shifting to a bio-economy risks diverting biological resources to purposes other than food, without necessarily significantly reducing GHG emissions.

There are two main areas where a better alignment of policies would allow a bio-economy to really contribute to the challenge of food security and the climate agenda.

First, the key issue is how to ensure the sustainability of the entire value chain, including primary production and inputs, logistics and transport, the conversion of the bio-resources into multiple products, and finally the distribution, sale and end-use of the product. Whether a bio-economy helps or hinders a low-carbon and sustainable future depends on rigorous accounting of the carbon flows along the whole chain. Governments must quantify emissions at each stage and compare them to alternative sources. Research has shown that GHG emissions over the entire life cycle of bio-products can be worse than for fossil fuels if the feedstock is grown on previously non-cultivated areas (OECD, 2014g). For instance, GHG emissions can arise either by growing the biomaterial on land that was previously forest, or indirectly by displacing the production of other commodities which are then produced elsewhere (IEA, 2012). There is still much uncertainty associated with the models meant to account for this latter effect, so more research is needed to allow for more rigorous accounting of these emissions. Sourcing biomaterials from recycled products or food waste could avoid those emissions.

Second, the synergies and trade-offs between a bio-economy and other goals, such as food security, biodiversity conservation and resource management, need to be properly assessed.

Making sure that a bio-economy approach delivers on all these crucial areas will require aligned policy signals and the following elements (IEA, 2012):

- Internationally aligned certification schemes that promote the sustainable production of energy crops, ensure GHG benefits and avoid competition for land with growing food demand (Box 9.8).
- Considerable investment in agriculture: For example, investment is needed in feedstock and production R&D to accelerate the development of technologies that do not interfere with food production.
- More open biomass and biofuels trade: Trade will play a crucial role in matching supply and demand in different regions, calling for a reduction of existing tariffs on biomass products. Some countries' import tariffs, especially for ethanol, remain very high (greater than 30%) and should be reduced or eliminated. Internationally agreed technical standards for biomass and biomass intermediates would remove an important current impediment to trade.

Box 9.8. Biofuels and food prices

In 2007-08, major food staples saw their biggest increase in prices since the 1970s. Many blamed the impact of biofuels policies, which they believed were encouraging the growth of biofuels instead of food. A number of studies and experts attempted to quantify the real impact of biofuel mandates on the 2007 food crisis, but were unable to reach a consensus. For instance, in 2010 the World Bank revisited a statement published in 2008 claiming that biofuels were responsible for 70% of the price increase (Mitchell, 2008). The later research found that trade policies were likely to be the more significant driver (World Bank, 2010).

Recent studies assessing plausible futures for agricultural markets, large-scale bioenergy and biofuels use and global food security are also contradictory. Lotze-Campen et al. (2014) used five agro-economic models to analyse the impact on food prices of a scenario in which global bioenergy would be consistent with a 2°C pathway. The authors estimate that the global impact on food prices of "an ambitious mitigation scenario with high bioenergy demand" (+5% average across models) would be much lower than price impacts of climate change on crop yields in a high-emissions scenario (25% average across models).

- Biofuel support policies that are conditional on their life-cycle GHG performance: Many countries are increasingly mainstreaming sustainability criteria into their biofuels policy and practices. For example, under its Renewable Energy Directive, the EU demands that sustainability criteria for biofuels be met to be eligible to satisfy binding national targets. The criteria include at least 35% GHG emissions savings compared with fossil fuels – this will rise to 50% in 2017 and 60% in 2018 for biofuels produced from new facilities. The United States Environmental Protection Agency's Renewable Fuel Standard II programme states that advanced biofuels must demonstrate that they meet minimum GHG reduction standards of 50% and 60% along the entire life cycle.
- Biofuel environmental impact assessment which accounts for emissions from indirect land-use changes: Some governments have already tried to account for indirect land-use change. For example, the state of California's low-carbon fuel standard adds an overall GHG penalty to calculations to account for them. Other policies to reduce the impact of land-use change include zoning policies. For example, Brazil's Agro-Ecological Sugarcane Zoning Programme limits the areas in which sugar cane production can be expanded. The programme limits access to development funds for non-compliers (IEA, 2011).
- A focus on using waste and residues, perennial energy crops and maximising land-use efficiency through improved productivity or the use of specific high-yielding feedstock in order to avoid emissions from biofuel-driven land-use change.

Beyond bioenergy, many technical barriers also stand in the way of the deployment of a bio-based economy. For example, complex and time-consuming regulations can drastically affect the small bio companies that are so often sources of innovation (Box 9.9). Governments should undertake an assessment of their regulatory framework and more develop appropriate tools and methodologies to make sure the bio-economy can support growth while preserving climate, food security and a broader development agenda.

Box 9.9. Barriers to a bio-economy

The fast pace at which bio-chemicals have developed has made it hard for governance and regulatory frameworks to keep up. Though the idea of "bio-refineries" – locations which group the production of fuels, chemicals and plastics from bio-feedstock – is gaining ground, decades of research in bio-chemicals have not led to commercial progress. One problem is that bio-based production has not received much policy attention and is undermined by complex and time-consuming regulations that have a drastic impact on small bio companies.

A study for the government of the Netherlands (Sira Consulting, 2011) identified around 80 regulatory barriers to the bio-based economy.

- Fundamental constraints: These call for a political and policy approach (e.g. import duties, level playing field, certification and financial feasibility).
- Conflicting objectives: These barriers cannot be removed, but governments can help companies to meet the regulations (e.g. REACH regulations).
- Structural constraints, requiring adjustments to regulations: These require adjustment to regulations, but do not demand policy or political action.
- Operational constraints: Here the regulations themselves are not the problem but rather their implementation, for example by local authorities. This leads to substantial barriers to investment in the bio-economy, especially for small and medium-sized enterprises.

Source: Extracted from OECD (2014g), "Biobased chemicals and bioplastics: Finding the right policy balance", *OECD Science, Technology and Industry Policy Papers*, No. 17, OECD Publishing, Paris, http://dx.doi.org/10.1787/5jxwwfjx0djf-en based on Sira Consulting (2011), Botsende belangen in de biobased economy. Een inventarisatie en een analyse van de belemmeringen in de transitie naar een biobased economy, Sira Consulting, The Hague.

Waste and over-consumption need to be tackled

One of the final pieces of the low-carbon land-use puzzle is how to reduce losses and waste of food and other products derived from the land (more broadly, any type of waste derived from land). The FAO defines food loss as "the decrease in the quantity or quality of food," and food waste as the part of food loss which is fit for consumption but removed from the food supply chain "by choice, or which has been left to spoil or expire as a result of negligence by the actor – predominantly, but not exclusively the final consumer at household level." While the term "food loss" encompasses food waste, the FAO uses the term food loss and waste "to emphasi[s]e the importance and uniqueness of the waste part of food loss" (FAO, 2014b).The FAO estimates that roughly one-third of the food produced globally for human consumption is lost or wasted every year (FAO, 2009).

The worldwide loss and waste of food could have a significant impact on GHG emissions, by increasing the need for more land-clearing, more fertiliser and pesticide applications, more fuel used in farm equipment and to bring the food to market, and more methane emitted from rotting food in landfills. The FAO (2013) estimates that 3.3 billion tonnes of GHG are added to the atmosphere from food that is produced but not eaten, making it the third-highest emitter after the United States and China. Food losses and waste also imply a number of other economic, environmental and social impacts, such as the inefficient use of resources and inefficient supply chains for agricultural products. The direct economic cost of food wastage could add up to USD 650 billion a year (FAO, 2013).

The OECD's Green Growth Strategy targets reducing food waste as a means to increase food supply and reduce pressures on land (OECD, 2013d). But doing so means identifying and addressing the root causes. In developing and low-income countries, food waste and losses mainly occur at the early stages of the food supply chain through inefficient harvesting techniques, storage and cooling facilities, transport, packaging and marketing systems (Table 9.2). Addressing these inefficiencies will require investments in infrastructure, and technical and financial support to farmers and along the supply chain.

Table 9.2. **Food loss and waste along the supply chain**

Production	Handling and Storage	Processing and Packaging	Distribution and Market	Consumption
Definition				
During or immediately after harvesting on the farm	After produce leaves the farm for handling, storage and transport	During industrial or domestic processing and/or packaging	During distribution to markets, including losses at wholesale and retail markets	Losses in the home or business of the consumer, including restaurants/caterers
Includes				
Fruits bruised during picking or threshing	Edible food eaten by pests	Milk spilled during pasteurisation and processing	Edible produce sorted out due to quality	Edible products sorted out due to quality
Crops sorted out post-harvest for not meeting quality standards	Edible produce degraded by fungus or disease	Edible fruit or grains sorted out as not suitable for processing	Edible products expired before being purchased	Food purchased but not eaten
Crops left behind in fields due to poor mechanical harvesting or sharp drops in prices	Livestock death during transport to slaughter or not accepted for slaughter	Livestock trimming during slaughtering and industrial processing	Edible products spilled or damaged in market	Food cooked but not eaten

Sources: Extracted from Bagherzadeh, M. et al. (2014), "Food waste along the food chain", *OECD Food, Agriculture and Fisheries Papers*, No. 71, OECD Publishing, Paris, http://dx.doi.org/10.1787/5jxrcmftzj36-en, based on Lipinski, B. et al. (2013), "Reducing food loss and waste", Installment 2 of "Creating a Sustainable Food Future", Working Paper, World Resources Institute, Washington, DC, available at: www.wri.org/sites/default/files/reducing_food_loss_and_waste.pdf.

Losses at the consumer level, both household and out-of-home eating, are high and growing in urban areas of developing countries. In OECD countries, it seems that food is mainly wasted at later stages of the supply chain; raising awareness and consumer engagement are key. Governments can also improve collaboration between different segments of the value chain through a better alignment of different policies. For instance, the Good Samaritan Act in the United States and similar legislation in several EU countries protect donors from liability, hence encouraging donation to food banks. Other countries send contradictory signals by exempting discarded food from value-added tax but not donated food (Bagherzadeh et al., 2014).

Several areas require a better alignment of policies (OECD, 2013a):

- Improve data and reporting: The availability of household data on food waste is relatively good in OECD countries. However, there is very little national information on food waste in other parts of the food chain, particularly manufacturing, distribution and out-of-home eating. Adopting common definitions of food waste and developing accurate reporting of food waste by economic activity and commodity groupings would contribute to a better understanding of the challenge and allow tailored recommendations (OECD, 2013a). As the cost of food loss and waste data collection could be very high, a systematic analysis of the food supply chain to identify the major drivers of food loss and waste could provide actionable information in the absence of quantity data.

- Improve the legal and institutional framework for food waste: Most countries have a legal framework for food waste, but it is usually embedded in a broader waste framework under the authority of the Ministry of Environment or local governments and municipalities responsible for waste collection management and recycling services. Food waste is only implicitly included in general overarching objectives relating to waste, such as avoiding the generation of waste, managing waste as a resource, ensuring safe and environmentally friendly waste treatments, and efficient waste management. The legal and institutional setting does not favour co-ordination between different ministries and does not address the root causes of waste: the Ministry of Agriculture or Economy is in charge of farming activities and food industry oversight, while the Minister of Environment regulates waste management, the Ministry of Economy administers tax incentives for food donation and the Ministry of Health oversees the safety aspects of food, usually through a Food Safety Authority under its control. This fragmented approach to food waste creates potential conflicts of interests between different authorities and government bodies and is hindering the transition to a more efficient management of the food supply chain.
- Strengthen collaboration with the private sector: The government has an important role to play in avoiding food waste while providing safe food. However, the private sector can influence many aspects of the management of the value chain, from innovation in production processes to a more informed use of labels. For instance, the "best before" label does not come as a recommendation from governments, but is an industry practice to adapt to expected business liabilities or maximise taste and flavour. National governments should provide the right policy framework and incentives to facilitate partnerships between industries and raise awareness; and develop education campaigns. Several platforms for co-operation have been implemented to facilitate this dialogue, as part of the FAO's Save Food initiative.[5]

Towards an integrated and holistic approach

There are many synergies – and some trade-offs – between the climate agenda and the pursuit of other environmental goals, such as reducing water pollution and conserving biodiversity. A number of policy instruments need to be better aligned to achieve a low-carbon, resilient economy without compromising other fundamental policy goals.

International, national and local policies can create synergies between the different agendas if they explicitly factor in multiple goals in policy formulation. For instance, at the international level, some policies could promote synergies between reduced deforestation and enhanced agricultural production, including: agricultural soil carbon credits in carbon markets, strengthening development assistance for projects pursuing the three objectives, developing guidelines for the participation in global value chains that encourage the production of commodities on land that has already been cleared (DeFries and Rosenzweig, 2010). More work is needed to identify how policy makers could help manage synergies and trade-offs between the different agendas, for a more inclusive green growth (OECD, 2014a).

Notes

1. Environmental cross-compliance criteria aim to address several environmental objectives, including reduction of nitrogen and phosphorus runoff and leaching, soil erosion and sediment runoff, conservation and promotion of farmland biodiversity, reduction of pesticide runoff and residues, and improved animal welfare.

2. See: www.globalresearchalliance.org.

3. For details see www.un-redd.org/aboutredd.

4. UNFCCC's Warsaw Framework on REDD+.

5. See: www.save-food.org.

References

Alexandratos, N. and J. Bruinsma (2012), "World agriculture towards 2030/2050: The 2012 revision", *ESA Working Paper*, No. 12-03, Food and Agriculture Organization, Rome, June, available at: www.fao.org/docrep/016/ap106e/ap106e.pdf.

Bagherzadeh, M. et al. (2014), "Food waste along the food chain", *OECD Food, Agriculture and Fisheries Papers*, No. 71, OECD Publishing, Paris, http://dx.doi.org/10.1787/5jxrcmftzj36-en.

Brooks, J. (2014), "Policy coherence and food security: The effects of OECD countries' agricultural policies", *Food Policy*, Vol. 44, pp. 88-94, February, http://dx.doi.org/10.1016/j.foodpol.2013.10.006.

Brooks, J.(ed.) (2012), *Agricultural Policies for Poverty Reduction*, OECD Publishing, Paris, http://dx.doi.org/10.1787/9789264112902-en Brooks, J. and A. Matthews (2015), "Trade dimensions of food security", *OECD Food, Agriculture and Fisheries Papers*, No. 77, OECD Publishing, Paris, http://dx.doi.org/10.1787/5js65xn790nv-en.

Clark, A.J. et al. (2012), *Impacts of Climate Change on Land-based Sectors and Adaptation Options*, MPI Technical Paper, No. 2012/33, Ministry for Primary Industries, Wellington, New Zealand, November, available at: www.mpi.govt.nz/document-vault/4054.

Climate Summit (2014), "Global Alliance for Climate-Smart Agriculture: Action plan", available at: www.un.org/climatechange/summit/wp-content/uploads/sites/2/2014/09/AGRICULTURE-Action-Plan.pdf.

DeFries, R. and C. Rosenzweig (2010), "Toward a whole-landscape approach for sustainable land use in the tropics", *Proceedings of the National Academies of Science*, Vol. 107, pp. 19 627-19 632, http://dx.doi.org/10.1073/pnas.1011163107.

EC (2013), *European Union Timber Regulation (EUTR) 2013*, background briefing for media, European Commission, Brussels, available at: http://ec.europa.eu/environment/eutr2013/_static/files/press-briefings/eu-timber-regulation-media-briefing_v01_en.pdf.

Ecosystem Marketplace (2014), "Turning over a new leaf: State of the forest carbon markets 2014", Forest Trends' Ecosystem Marketplace, Washington, DC, available at: www.forest-trends.org/documents/files/doc_4770.pdf

EU (2013), Guidance Document for the EU Timber Regulation, September, available at: http://ec.europa.eu/environment/forests/pdf/Final%20Guidance%20document.pdf.

FAO (2014a), "Agriculture, forestry and other land use emissions by sources and removals by sinks: 1990-2011 analysis", *FAO Statistics Division Working Paper Series*, ESS/14-02, Food and Agriculture Organization, Rome, March, available at: www.fao.org/docrep/019/i3671e/i3671e.pdf.

FAO (2014b), "Definitional framework of food loss", Working Paper, Food and Agriculture Organization, Rome, 27 February, available at: www.fao.org/fileadmin/user_upload/save-food/PDF/FLW_Definition_and_Scope_2014.pdf.

FAO (2014c), Principles for Responsible Investment in Agriculture and Food Systems, Food and Agriculture Organization, Rome, www.fao.org/3/a-ml291e.pdf.

FAO (2013), *Food Wastage Footprint: Impacts on Natural Resources: Summary Report*, Food and Agriculture Organization, Rome, available at: www.fao.org/docrep/018/i3347e/i3347e.pdf.

FAO (2012), *Voluntary Guidelines on Responsible Governance of Tenure of Land, Fisheries and Forests in the Context of National Food Security*, Food and Agriculture Organization, Rome, available at: www.fao.org/docrep/016/i2801e/i2801e.pdf.

FAO (2009), "How to feed the world in 2050. Issues brief for the high-level forum on How to Feed the World in 2050", Food and Agriculture Organization, Rome.

FAO (2002), *World Agriculture: Towards 2015/2030*, Food and Agriculture Organization, Rome, available at: www.fao.org/docrep/004/y3557e/y3557e00.htm.

Fishbein, G. and D. Lee (2015), "Early lessons from jurisdictional REDD+ and low emissions development programs", The World Bank and The Nature Conservatory, Rep. Arlington: n.p., available at: www.nature.org/media/climatechange/REDD+_LED_Programs.pdf.

GCEC (2014), *Better Growth, Better Climate: The New Climate Economy Report*, The Global Commission on the Economy and Climate, Washington, DC, October, available at: http://newclimateeconomy.report/wp-content/uploads/2014/08/GCEC_GlobalReport.pdf.

Gibbs, H.K. et al. (2009), "Tropical forests were the primary sources of new agricultural land in the 1980s and 1990s", *Proceedings of the National Academy of Sciences*, Vol. 107, No. 38, pp. 16 732-16 737, http://dx.doi.org/10.1073/pnas.0910275107.

IEA (2014), *Energy Technology Perspectives 2014*, International Energy Agency, Paris, http://dx.doi.org/10.1787/energy_tech-2014-en.

IEA (2012), *Bioenergy for Heat and Power*, IEA Technology Roadmaps, International Energy Agency, Paris, http://dx.doi.org/10.1787/9789264123236-en.

IEA (2011), *Biofuels for Transport*, IEA Technology Roadmaps, OECD Publishing, Paris, http://dx.doi.org/10.1787/9789264118461-en.

Ignaciuk, A. and D. Mason-D'Croz (2014), "Modelling adaptation to climate change in agriculture", *OECD Food, Agriculture and Fisheries Papers*, No. 70, OECD Publishing, Paris, http://dx.doi.org/10.1787/5jxrclljnbxq-en.

IPCC (2014), Intergovernmental Panel on Climate Change, Working Group III, Summary for Policymakers, www.ipcc.ch.

Lambin, E.F. et al. (2014), "Effectiveness and synergies of policy instruments for land use governance in tropical regions", *Global Environmental Change*, Vol. 28(2014), pp. 129-140, September, http://dx.doi.org/10.1016/j.gloenvcha.2014.06.007.

Lipinski, B. et al. (2013), "Reducing food loss and waste", Installment 2 of "Creating a Sustainable Food Future", Working Paper, World Resources Institute, Washington, DC, available at: www.wri.org/sites/default/files/reducing_food_loss_and_waste.pdf.

Lotze-Campen, H. et al. (2014), "Impacts of increased bioenergy demand non global food markets: An AgMIP economic model intercomparison", *Agricultural Economics*, Vol. 45, Issue 1, pp. 103-116, January, http://dx.doi.org/10.1111/agec.12092.

Meyfroidt, P. and E. Lambin (2011), "Global forest transition: Prospects for an end to deforestation", *Annual Review of Environment and Resources*, Vol. 36, pp. 343-371, http://dx.doi.org/10.1146/annurev-environ-090710-143732.

Millenium Ecosystem Assessment (2005), *Ecosystems and Human Wellbeing: Current State and Trends*, Island Press, Washington, DC.

Mitchell, D. (2008), "A note on rising food prices", *World Bank Policy Research Working Paper Series*, No. 4682, World Bank, Washington, DC, July, http://dx.doi.org/10.1596/1813-9450-4682.

OECD (2015a), "The role of public policies in promoting adaptation in agriculture", COM/TAD/CA/ENV/EPOC(2014)13/FINAL, OECD, Paris, available at: www.oecd.org/officialdocuments/publicdisplaydocumentpdf/?cote=COM/TAD/CA/ENV/EPOC%282014%2913/FINAL&docLanguage=En.

OECD (2015b), "A review of literature on the cost-effectiveness of greenhouse gas mitigation measures for agriculture", OECD, Paris.

OECD (2015c), *Fostering Green Growth in Agriculture: The Role of Advisory Services, Training and Extension Initiatives*, OECD Green Growth Studies, OECD Publishing, Paris, http://dx.doi.org/10.1787/9789264232198-en.

OECD (2015d), *Policy Approaches to Droughts and Floods in Agriculture*, OECD Publishing, Paris, forthcoming.

OECD (2014a), "Synergies and trade-offs between adaptation, mitigation and agricultural productivity: A scoping paper", prepared for the 38th session of the Joint Working Party on Agriculture and the Environment, OECD, Paris.

OECD (2014b), *Green Growth Indicators 2014*, OECD Green Growth Studies, OECD Publishing, Paris, http://dx.doi.org/10.1787/9789264202030-en.

OECD (2014c), *Climate Change, Water and Agriculture: Towards Resilient Systems*, OECD Studies on Water, OECD Publishing, Paris, http://dx.doi.org/10.1787/9789264209138-en.

OECD (2014d), *Agricultural Policy Monitoring and Evaluation 2014: OECD Countries*, OECD Publishing, Paris, http://dx.doi.org/10.1787/agr_pol-2014-en.

OECD (2014e), *Towards Green Growth in Southeast Asia*, OECD Green Growth Studies, OECD Publishing, Paris, http://dx.doi.org/10.1787/9789264224100-en.

OECD (2014f), *Policy Framework for Investment in Agriculture*, OECD Publishing, Paris, http://dx.doi.org/10.1787/9789264212725-en.

OECD (2014g), "Biobased chemicals and bioplastics: Finding the right policy balance", *OECD Science, Technology and Industry Policy Papers*, No. 17, OECD Publishing, Paris, http://dx.doi.org/10.1787/5jxwwfjx0djf-en.

OECD (2014h), *Green Growth Indicators for Agriculture: A Preliminary Assessment*, OECD Green Growth Studies, OECD Publishing, Paris, http://dx.doi.org/10.1787/9789264223202-en.

OECD (2013a), *Policy Instruments to Support Green Growth in Agriculture: A Synthesis of Country Experiences*, OECD Publishing, Paris, http://dx.doi.org/10.1787/9789264203525-en.

OECD (2013b), *OECD Compendium of Agri-environmental Indicators*, OECD Publishing, Paris, http://dx.doi.org/10.1787/9789264186217-en.

OECD (2013c), *Agricultural Policy Monitoring and Evaluation 2013: OECD Countries and Emerging Economies*, OECD Publishing, Paris, http://dx.doi.org/10.1787/agr_pol-2013-en.

OECD (2013d), *Global Food Security: Challenges for the Food and Agricultural System*, OECD Publishing, Paris, http://dx.doi.org/10.1787/9789264195363-en.

OECD (2012), *OECD Environmental Outlook to 2050: The Consequences of Inaction*, OECD Publishing, Paris, http://dx.doi.org/10.1787/9789264122246-en.

OECD (2011), *Managing Risk in Agriculture: Policy Assessment and Design*, OECD Publishing, Paris, http://dx.doi.org/10.1787/9789264116146-en.

OECD (2010), *Paying for Biodiversity: Enhancing the Cost-Effectiveness of Payments for Ecosystem Services*, OECD Publishing, Paris, http://dx.doi.org/10.1787/9789264090279-en.

Ostrom, E. (1990), *Governing the Commons: The Evolution of Institutions for Collective Action*, Cambridge University Press.

Sayer, J. et al. (2013), "Ten principles for a landscape approach to reconciling agriculture, conservation, and other competing land uses", *Proceedings of the National Academy of Sciences*, Vol. 110, No. 21, pp. 8 349-8 356, http://dx.doi.org/10.1073/pnas.1210595110.

Sira Consulting (2011), *Botsende belangen in de biobased economy. Een inventarisatie en een analyse van de belemmeringen in de transitie naar een biobased economy*, Sira Consulting, The Hague.

Von Lampe, M. et al. (2014), "Fertiliser and biofuel policies in the global agricultural supply chain: Implications for agricultural markets and farm incomes", *OECD Food,*

Agriculture and Fisheries Papers, No. 69, OECD Publishing, Paris, http://dx.doi.org/10.1787/5jxsr7tt3qf4-en.

Wiggins, S. and J. Brooks (2010), "The use of input subsidies in developing countries", TAD/CA/APM/WP(2010)45, OECD, Paris, available at: www.oecd.org/tad/agricultural-policies/46340359.pdf.

World Bank (2013), *World Development Indicators 2013*, World Bank, Washington, DC, http://dx.doi.org/10.1596/978-0-8213-9824-1.

World Bank (2010), "Placing the 2006-08 commodity price spike into perspective", *Policy Research Working Paper*, World Bank, Washington, DC.

World Food Summit (1996), *Rome Declaration on World Food Security*, Food and Agriculture Organization Publishing, Rome, available at: www.fao.org/docrep/003/w3613e/w3613e00.HTM.

Zhang, W. et al. (2013), "New technologies reduce GHG emissions from nitrogenous fertilizer in China", *Proceedings of the National Academy of Sciences*, Vol. 110, No. 21, pp. 8375-8380, http://dx.doi.org/10.1073/pnas.1210447110.

OECD PUBLISHING, 2, rue André-Pascal, 75775 PARIS CEDEX 16
(03 2015 02 1 P) ISBN 978-92-64-23326-3 – 2015-02

www.ingramcontent.com/pod-product-compliance
Ingram Content Group UK Ltd.
Pitfield, Milton Keynes, MK11 3LW, UK
UKHW051500290926
15471UKWH00008B/1314

9 789264 233263